Jacob Ebbtide

Introduction

While the mainstream media tends to downplay worries over resources such as petroleum the reality is that humanity is facing a crisis.

It is axiomatic that bad news is bad for business. It is not surprising that major news outlets downplay or simply fail to discuss issues of energy resources. The failure of the newsgathering and distribution networks does not negate reality. We are facing an energy crisis

Years ago there was talk of a *soft landing*. In talking about a 'soft landing' one of the possibilities suggested was developing new plant varieties such as switchgrass or other fast-growing plants to use as biomass for the conversion to biofuel. That is the basic thesis of this work.

The deserts are there. The atmosphere is about eighty percent nitrogen. There are bacteria that take nitrogen out of the air and symbolically provide nitrogen to plants.

There are also ways of 'wringing' moisture out of the air—rather than continue to deplete the earth's aquifers. There are earth-friendly ways of constructing environmental water resources. *Railroads, Biomass and Synthetic Oil,* a booklet by the same author, illustrates some of these techniques.

The narrative in *Jacob Ebbtide* makes the huge assumption that the governments of the world, acting through research universities, actually decide to put funding where it would do some good.

Expecting governments or corporate America to put money into large-scale 'energy-agriculture' for the growing of biomass in areas not suited for conventional food production is, of course, pure fantasy. In that sense this work is fiction, but in its underlying assumption that we face an energy crisis is fact.

1

Late one evening Doctor Jacob Ebbtide, fortyish, slim with gray around his temples, walked through his laboratory at Tel Aviv University.

Only yesterday he'd received written notice of his house arrest. It came from the Office of President Ben Leviman. It did, however, given him permission to continue work in his laboratory at Tel Aviv University.

A computer was turned on and sitting on a workbench. The computer hadn't been there an hour ago. On the computer he found a neatly typed note with an Internet address. The note also had not been there. Jacob keyed in the Internet address and silently watched the video of three men tied to wooden posts with rope. Shots rang out. The men fell. The camera zoomed in for a close shot at each man hanging from a post.

There was no voice message and only a caption on the screen: 'Traitors.' The video ended. Jacob switched off the computer.

Jacob was under house arrest. He didn't know why. All communication between him and the world had been severed. No phone calls, no emails and no faxes. He was under arrest in his own home and laboratory, but no formal charges had been made. The authorities were not talking.

Jacob felt that his work as the lead scientist at the Agricultural Genetics Laboratory was undoubtedly at the root of his troubles with the Israeli legal system.

Two of his most trusted friends and colleagues Herman Burger and Emil Zada, both senior scientists, were missing. They had vanished two weeks ago.

His own daughter Talya had also vanished at the same time. He knew he was in trouble, but what could he do about it?

Outside a blistering hot day gave way to a muggy star-filled night. Walking out of a greenhouse onto the roof Jacob stood watching distant flashes of artillery exchanges between the Israelis and a hodgepodge of militant groups in Lebanon.

The dusty scent of cordite drifted miles in hot smothering air. With the resumption of the attacks there was a good chance that the security forces encamped around the University would have their backs turned to watch the distant flashes of gunfire as they gossiped. It might give Jacob a way to get out to the growing fields.

He was the scientist responsible for rushing a very special crop into production. The crop was to alleviate Israeli's chronic liquid fuel energy shortages.

The answer to his legal troubles Jacob was sure was out in the university growing fields. He was under house arrest for a reason and that reason he felt sure had to do with the experimental crop.

Jacob believed he could find the reasons for his house arrest if he could see the condition of the fields. His troubles had to be centered in his work, because there was little in his life other than his research.

Tonight he needed a chance to slip away. He trusted his driver Masal. Leaning over a railing, Jacob looked down three floors at well-lighted walkways and courtyards around the Biological Sciences Building. Then over at a repair depot he saw Masal standing by a shed. He looked up. Jacob waved to him.

Masal motioned for Jacob to come down. This confused him, because guards were posted at the doors. When Masal made more aggressive hand motions Jacob went to the laboratory door and found no guards. The way was clear.

Jacob knew that Masal had retired from the Mossad. He now worked as Jacob's driver, but how much 'pull' Masal had with President Leviman or the Israeli Jacob had no way of knowing.

Fifteen minutes later Jacob stepped out of Marsal's old jeep and used a powerful searchlight on the vehicle to view a field of withering plants. Low, neat rows of Solbean, the new mainstay of Israeli energy-agriculture, spread out before him in one of many University growing plots.

Solbean was grown as forage. Livestock thrived on it. People also put it in their salads and stews. It was a variety of soybean with a leguminous herbaceous growth habit.

It's primary use, however, had never been to feed livestock or for food. The plant was so hardy and cheap to grow that it was distilled into alcohol and used as fuel.

It ran everything in Israel and in many other oil-starved regions of the world from farm tractors to seawater distillation plants.

"Other fields?" Jacob asked.

"The same," Masal said.

Taking a flashlight from Masal and walking into the field, knelling down and tearing off a leaf, Jacob saw the problem. From leaf color the plant was dying from drought; yet, dark stains trickled down through poor sandy soil. Dark patches of moisture darkened the soil as the drip irrigation system released precious water. Solbean, a miracle plant, fixated nitrogen directly out of the air and promised energy to an energy-starved country.

"Can it be saved?" Masal came into the field.

"Call Dexter at the water plant. Tell him to turn off the irrigation."

Jacob held the leaf up to the flashlight. Swollen dark purple veins in the leaf told him that the plant had water, but it was dying nonetheless. Jacob knew the answer. In test plots in his laboratory Jacob had seen that distinct type of colorization. Solbean leaves turned pale blue, almost white, prior to death. The field had maybe two days to be saved.

Jacob knew that the plant could not metabolize nutrients into carbohydrates, because it wasn't getting nitrogen. It is nitrogen that gives leaves their green color.

"It's like this all over the country?" Jacob asked.

"From Har Meron to Beersheba." Masal hesitated and then added, "Leviman executed three spies today."

"I saw that on the Internet," Jacob said.

He didn't tell Masal that he thought Leviman's agents had put a threatening note in his laboratory.

"He had them shot," Masal added. "I heard that they were working for the Americans. They were engineers."

"What crime?"

"They were charged with espionage."

"Spying?" Jacob asked. "On us? What for?"

"Don't know."

"What do you hear?"

"They hacked into computer networks," Masal said.

Masal walked out of the field and stood by the jeep. When Jacob came up to the road he saw his driver's face. Masal was troubled.

"The crop has been dying for a couple days," Masal said "Leviman is out for blood."

"A couple days?"

Now the timing of his house arrest made sense. Jacob suspected his trouble had something to do with the Solbean crop, but that the whole crop was in danger never crossed his mind. The sheer scale of the disaster made Jacob blink tears.

"I should have been notified!" he shouted. "House arrest or not!"

Masal faced him. "We have orders! If Leviman wanted you to know he'd have the guards bring you out here."

"And you?" Jacob asked. "You help me? Why?"

"I know you," Masal said. "It's that simple. You're not a traitor. I know it, but I can't get others to agree. They see the crop and think of you."

Jacob mopped sweat off his face with his shirtsleeve. He heard the drone of an aircraft. He knelt down by the jeep to shelter from a helicopter flying low over the dirt road. Its searchlight swung side to side as if searching for a fugitive hiding in the fields.

Jacob squatted as the helicopter flew over and then stood up and used the vehicle's searchlight to let his gaze take in the whole field.

It was obvious to his practiced eye that this was no accident. The condition of the crop was not incompetence. This was sabotage. He knew it and so did Leviman. No wonder Leviman was shooting people.

"About the crop?" Masal asked. "Can it be saved?"

"The crop isn't getting nitrogen," Jacob said. The connection between the condition of the crop and his house arrest hit him again. It was obvious. He was the lead scientist and the crop was dying. Those dots were easy to connect.

"Can it be saved?" Masal asked again.

"I don't know."

Even a week ago Jacob had noticed people on campus that he believed were Israeli intelligence operatives. That would not be unusual, but he had gotten word that he was the subject of their inquiries.

Somebody, Jacob thought, was planning something. The presence of agents asking questions and taking notes gave Jacob a mild case of the nerves. Still Israeli was at war and having agents around was more or less a routine.

Then two weeks ago his two most important colleagues had also disappeared. Herman Burger and Emil Zada were both senior researchers.

So had his daughter. At first Jacob suspected kidnapping, because it was big business in the modern world. Kidnapping was common.

Businessmen and scientists were the priority targets, but even workmen were abducted in broad daylight and sometimes taken by force out of a crowd. Senior people such as Jacob often hired retired Mossad people such as Masal as drivers or security.

No ransom demands had been made for Herman or Emil or for Talya. Although criminal gangs were common Jacob suspected Leviman. Jacob believed President Ben Leviman was behind the abductions, because without demands for money their disappearances must have been political.

Somebody with authority was pulling strings. The note on the computer only added to his conviction that Leviman was behind everything—including the possibility that the President of Israel was destroying the Solbean crop.

Having top scientists disappear was a professional problem for Jacob, but losing the crop was disaster for Israel. And, it was a guaranteed firing squad for anybody even suspected of having anything to do with the loss of the crop.

"Can it be saved?" Masal asked again.

Jacob heard Marsal's question, but his busy mind was looking for answers. The crop wasn't metabolizing nitrogen and that in Jacob's thinking also brought the Americans into the picture. The Americans had paid for the development of Solbean. It was their intellectual property.

Jacob looked to the dark eastern sky. Smoke and haze drifted in from Lebanon. The smoke seemed symbolic of the war shaping up between Israel and the Americans and the Arabs. This war was for survival. Energy was the weapon of choice. The core of the weapon was the genetic Code for Solbean.

The Americans, along with their Israeli contractors, had developed Solbean at Tel Aviv University, but the Americans paid for it. They owned it.

The Americans had the rights to Solbean, but given that a desperate world was less and less likely to respect patent rights the Americans had made a genuine effort to protect their investment by genetically engineering a special bacterium that fixated atmospheric nitrogen for the plant.

The Americans, through their Israeli contractors, had positively outdone themselves with the development of the Rhizobium living on the plant's roots.

Nitrogen fixation could only take place in the absence of oxygen so the Rhizobium grew under the soil. The Americans developed a strain of Rhizobium that was sensitive to certain frequencies of microwave radiation that could penetrate the soil for a few centimeters and reach the plant's shallow roots. The Rhizobium could detect ground-penetrating radiation the equivalent of a candle at five kilometers.

The Rhizobium had been developed in an effort to insure that Americans kept control of the Solbean and their means of maintaining control of the crop was to mount microwave transmitters on satellites.

Two satellites came over the horizon every night. They appeared at nine P.M. every evening with their antenna pointed down to the earth. Radiation would stop fixing nitrogen for the Solbean if the satellite swept overhead *without* radiating the crops. The radiation was a trigger that started the metabolic process of nitrogen fixation.

Without the microwave stimulus provided by the satellites the Solbean did not fixate nitrogen. The crop died. The Americans had genetically engineered Solbean and the Rhizobium such that neither the plant nor the bacteria could live without the other. It was a life and death symbiotic relationship.

The satellites could be active or inactive, but even if they were broadcasting microwave radiation that did not necessarily mean that the Solbean would live.

The Americans could change the frequency of the microwave radiation and if they did change the frequency the crop would die. Jacob looked up at the night sky. The satellites were probably overhead, but were they stimulating the crop or killing it?

And, if the Americans were killing the crop Jacob thought he knew why; they no longer trusted their Israeli partners. Making matters worse Israel might have given them reason to think they were getting betrayed.

Solbean was in development and testing, but the needs of Israel were so desperate that crops had been rushed into production in spite of American protests. So far the results had been wonderful.

In ten months two crops had been grown and processed into fuel. The plant's large green meaty leaves were the biomass of choice for the entire world. A mere thousand pounds of dried biomass from Solbean produced a barrel of high-grade liquid fuel.

As of two weeks ago heated messages between Tel Aviv and Washington had burned up the airways. Accusation and counter accusation had caused tempers to boil. Now it looked as if the Americans were pulling the plug.

"Get me back to the laboratory," Jacob said.

They drove along the fields. Jacob felt hot dust biting into his face; the whole planet was baking. In the fields the crops had water, but even that wouldn't last.

The water distillation plant processed seawater into drinking water, but even the distillation plant ran on alcohol distilled from Solbean. Without this crop all other agricultural crops and a large proportion of the population would go without water or food.

Jacob was sure it was sabotage. What else could it be? And who was behind it? Were the Americans finally putting their foot down? Were they taking action against Israeli preemptive production? From the condition of the leaves it looked that way.

Did Leviman think that people in his own research program were working with the Americans? If it wasn't the Americans why was the crop dying?

Jacob wondered if the Americans had the means to put intelligence operatives into his laboratory. What had Masal said? That the three spies worked in communications? That speculation made sense because all data is stored on computers.

Were the Americans hacking into University computers? If so, why? Jacob shook his head. He was sure Americans had all the data…or did they?

Maybe the Americans didn't trust the Israelis to give them the right data? Maybe the Americans were 'keeping tabs' on the Israelis? Maybe the Americans didn't trust their Israeli partners and were conducting their own espionage to make sure that the Israelis were not out to reprogram the Genetic Code?

"The Americans," Jacob realized that the satellite transmissions or the lack of the microwave stimulus would account for the condition of the crop. Was it the Americans?

"What?" Masal asked.

"The Americans," Jacob said again. "They have a mechanism for killing the crop. All plants need nitrogen. Solbean takes it out of the air with the help of a special bacterium. The Americans control the bacterium."

"Americans?" Masal asked.

"Yes," Jacob said. "We have a contract with them."

"The Americans? Why?"

"We rushed into production," Jacob said. "We broke the terms of the contract. The Americans paid hundreds of millions of gold dinars. They financed the program, but still it might not be the Americans. I'm not sure."

Masal brought the jeep to a halt, set the handbrake and looked at Jacob. "Is it the Americans?"

"The American's have no reason to kill our crops," Jacob said.

"But we broke our contract with them?" Masal asked.

"That's true," Jacob explained, "but we've been paying for each gallon for fuel produced. They are making money on their investment so I don't understand why they would take this kind of action. And besides we already grew two crops."

"But we broke the agreement?" Masal asked.

"We broke the agreement, but we're not stupid."

"What?"

"They have a lock on the plant," Jacob said. "We know it. We are paying them for our production. We are licensed partners with them. We are paying them," Jacob repeated.

"We are not cheating them! They can see from space how much we are growing."

"If not the Americans then who?" Masal asked.

"I don't know. I just don't know."

Masal straightened around in his seat. "America," he said, shaking his head. "Why would they do that?"

"It's money," Jacob said. "Lots of money. That's the problem. And there are the Arabs to think about."

"Arabs?"

"The Americans fear the Arabs, but you know many Americans still think they can deal their way out of problems. As for the Arabs I'm sure they see our work as a threat to their power. They have what's left of cheap oil. They see Solbean as a major competitor. It's all about price in a world gone to hell.

"The Americans don't want Arabs to have good breeding plants. For the Americans it's payback for the terrorist attacks they've suffered, but still Americans love making deals." Jacob's cell phone beeped.

"Doctor?" It was his lab assistant Maurice.

"Speaking."

"Are you safe?"

"Why wasn't I told about the Solbean? This is a disaster! I'm coming back to the lab. You be there!"

"Now listen, Doctor, I…"

"No," Jacob shouted, "You listen! You be there. In one hour!"

"It's President Leviman!" Maurice insisted. "He sent people to my office yesterday. They pushed me around! They had guns! They scared the hell out of me. They said that you are under house arrest. I was to cut off all communication between you and others outside the University!"

"It was you?" Jacob yelled.

"No outside calls," Maurice insisted. "No communication! I didn't know you had left the campus this evening. I'm scared. My family is scared!"

"Mossad?" Jacob asked.

"Yes," Maurice said. "They work for a special branch. They work for Leviman."

"Are you sure?"

"What do I tell them?" Maurice asked.

"Tell the truth. Say I left without you knowing."

"They think that you had something to do with the disappearance of Herman and Emil." After a short pause he asked, "Did you?"

The last two words made Jacob angry.

"Be there in an hour!" He snapped his cell phone closed. He turned to Masal.

"An hour?" Masal asked. "We're five minutes out."

"I need some private time."

"Private time?"

"I think Leviman is planning something for me."

"Scandal follows him," Masal said. "How could such a man become President of Israel?"

Jacob shrugged. "I'm a dual citizen of Israel and America. If things get bad here…"

Masal made a grunt of contempt turning his face away. Jacob looked over.

"I have a choice."

"Some choice."

They drove for a while. Masal asked, "What's Leviman got against you? I mean this seems personal. It's like he's out for blood."

"It is personal," Jacob said. "I got between him and my daughter."

"Ouch," Masal winced. "That's not good. Leviman has a reputation."

"He doesn't seem to care. Talya was only sixteen when he had an open affair with her."

"I remember that," Masal said. "Leviman controls the Mossad and he has an agenda."

Jacob felt grateful that he still had some connections in the Israeli secret service, although they didn't seem to be doing him much good these days. He might be able to reach out to some work-a-day operatives that he had known for years.

Jacob had worked for the Mossad. Without a history of involvement with the Israeli Secret Service Jacob knew he could be in much worse trouble.

While Jacob had a few connections such as Masal he knew Leviman was a rough man that had political power. It looked as if Leviman had sent people to the Campus to threaten staff and to keep Jacob ignorant of the problem with the crop.

Jacob was the lead scientist. He would take the fall if the crop failed. But why would Leviman do all that damage to control one scientist? Why would the President of Israel sabotage the crop?

Jacob rode with his head over on his chest, thinking. There was a history of mistrust and trouble between him and Leviman, but those personal problems were nothing compared with the entire world conspiring to control the Solbean crop. Whoever controlled Solbean controlled the world's new renewable energy resource.

Jacob looked skyward and then back at the dark landscape. The fields were getting sacrificed in an ongoing war between governments over who would control Solbean in a post-cheap petroleum world. How did Leviman fit into that conflict?

Forces were lining up on all sides and huge amounts of money were set to change hands. Three major players competed to control the Solbean crop, each with their own interests and agendas.

In the battle to control energy resources there were the Americans, the Israelis and the Arabs that were now mostly under the rule of the Kamsi religious sect.

"If you can tell me," Masal asked, his voice rising over the wind, "What's between you and Leviman? I mean, I know about Talya, but is there more?"

"I think Leviman wants a deal with the Kamsi Kingdom," Jacob said. "The Kamsi want the Solbean, but of course they need the technology to make it grow. It has vast potential. They want it for themselves."

"Leviman wouldn't betray Israel!"

"He may think it's better to beat the Americans to it," Jacob replied. "They love making deals. My money is on the Americans. As I said, maybe Leviman knows the Americans want a deal with the Arabs. Maybe he thinks is just a matter of time before the Americans breakdown and make a deal."

Jacob thought of a remark made many years before by the Soviet leader V.I. Lenin that American businessmen would elbow each other out of the way to sell the Soviets the rope by which they, the Soviets, would hang the Americans.

Still, when the Israeli contractors led by Jacob, Herman and Emil developed the Solbean Genetic Code they knew the stakes were high. They knew they would have to be clever at protecting the interests of the Americans that were paying a fortune for the development work, but also they would need to be clever to protect the interests of Israel. And so they put their heads together and came up with a plan.

There were, in fact, two different varieties of Solbean. The plant grown in the field that was subject to satellite stimulation was the first variety. It was the only variety known to the world.

A second variety would grow independently of the satellites, but only the Israelis knew of the other variety. Development of the second variety was absolutely necessary, because the Americans had contracted all of their satellites work to the Chinese and had done so for years. It was the Chinese that engineered the satellites and launched the rockets.

America, or what was left of it after the Breakup was completely dependent on the Chinese for technical support. The problem was that, once the Solbean was in common production and making money for the Americans the Chinese might decide to charge more money for their services; or, in the alternative refuse further technological support unless they were made partners.

China needed energy as much or more than most countries, because of China's enormous population and the needs of China's leaders to keeping their economy growing at any cost. And not just China.

India and Brazil and other countries were not far behind. They all needed energy and once Solbean was known and once its tremendous capability for producing liquid fuel was understood there would be all out war for each country to have its own independent source of fuel. Everybody was sick and tired of importing oil from other countries.

Oil had been a blessing for industrial growth, but had also been a curse. It put the receiving nations under terrible political and economic pressure. Aside from the endless wars fought over or about oil reserves there was the increasing cost of producing anything resembling conventional oil.

Costs were soaring. Many oil companies were either bankrupt or close to it. Those areas still producing oil at a price that people could afford was shrinking and yet, even with a declining population energy demand kept rising.

Both the Americans and the Israelis knew they had to find ways to keep control. And, the Solbean plant itself was a way to do that. The Israelis had worked hand in glove with the Americans to develop this phenomenal plant, but Jacob knew when push came to shove it was every country for itself.

Solbean was an agricultural miracle. It grew in poor soil or even in desert sand. It needed little water and little conventional fertilizer, because it took nitrogen directly out of the earth's atmosphere.

Jacob knew that the actual purpose of the first variety that needed a radiation stimulus to grow was to condition the soil. Solbean was *itself* a covert weapon—it poisoned the soil for other plants. In the process of fixating nitrogen from the atmosphere Solbean metabolized various molecular byproducts that made the nitrogen remaining in the soil not only unavailable to other crops, but also toxic to them.

The idea was to sign contracts with other countries and get them to plant the first variety of Solbean. Once they did those countries would be agriculturally captive to the Americans and the Israelis, because the only other plant that would grow would be Solbean. Jacob considered it a nasty business, but he had his orders.

Jacob, as the lead scientist on the project, also knew that the Americans worried that other countries could reuse plant seed. He knew the Americans didn't have to worry about that, because Solbean was self-terminating after one season. The seed was infertile. For additional crops other countries would have to come back to the Americans or the Israelis to buy seed.

This bare-knuckle approach was necessary, because the world had become a hard place after the terrorist attacks that did everything but completely destroy what was left of the Global Economy.

Other countries would undoubtedly try to develop their own varieties of Solbean. For that, however, they would need to break the Rhizobium Genetic Code. Even if they did, however, they would already be planting Solbean.

There was little chance of other countries actually breaking the Code. Unless a spy sold them the Code there was very little chance that the Rhizobium Code could be broken. It was the development of the Rhizobium Code that brought the Israelis into the program.

With the breakup of the United States it was understood that the Americans couldn't keep such a secret. The Israelis, however, are known for taking security seriously. It was Israeli security that persuaded the Americans to bring Israel into the program.

All of these considerations went through Jacob's head. As far as Jacob was concerned the politics of Solbean were as mind-bending as the Rhizobium Genetic Code.

As Jacob saw it there were two interrelated problems. The first was that the Americans had apparently decided that Israel was no longer a trustworthy partner.

There was nothing Jacob could do about that tonight. It was up to Leviman to bring the Americans back into the fold.

The second problem was Leviman. He was a known gangster that had through cunning and connivance stolen the last Presidential election.

Moreover, Leviman undoubtedly knew that to deal with the Arabs and get caught would mean a firing squad. If Leviman were dealing with the Arabs, making a preemptive deal before the Americans, he would have to make sure that somebody else took the fall. Jacob knew that as lead scientist on the Solbean project he was the likely choice.

"It's all about energy," Masal said.

His comment broke into Jacob's thinking. Jacob didn't answer, but lately wondered about the Americans. They were hard to figure. All the trouble they had made them unpredictable. What was once the United States was now in shambles with chunks of the previous country breaking off squabbling over energy resources.

First major pieces of the Interstate Highway System were sold off to foreign companies. Businessmen divided highways into toll roads. That was the first blow to American ego.

The public hated paying tolls when they were already paying two dollars and change for a British pint of gasoline. Jacob was surprised that the country hung together as long as it did, but when Maine seceded Connecticut was quick to follow.

Then the rush was on. The government squandered its military on endless wars for short-term profits so didn't have the resources to force wayward states back into the Union.

Jacob rode along thinking about the United Remnant and how fast the once powerful United States had split into numerous splinter republics and democracies.

"What about the Americans?" Masal asked.

"The Kamsi want a special deal with them. They want things the way they used to be. You know. When they had oil and Americans burned it like there was no tomorrow."

"What kind of deal?"

"The Kamsi want to grow Solbean for the American market. They figure that they can grow Solbean cheaper than Americans and drive them out of their own market

"Remember, the Americans even paid the Chinese to build the factories and finance Free Trade that undercut their own economy. You must remember, Masal, Americans still fervently believe in Free Trade. They confuse finance with economy. The lowest price will always attract them. They follow price like slaves. I think they learned from China, but anything is possible with the Americans."

2

It was a scorching summer evening toward the end of the decade when Dr. Donald Veeby made preparations to leave his job at the Georgetown University Radiation laboratory. He left work at six: thirty P.M.

His work required that he be on premises and not work from home. In his estimation the long string of virus infections that had ravaged the country was just another pestilence to add to the already long line of trouble.

As a nuclear chemist doing research with the Department of Energy on disposal methods for radioactive waste he simply forgot a small battery powered Geiger counter in his pants pocket. He slipped out of a white lab coat and into a jacket.

Driving along Connecticut Avenue he motored out of the District of Columbia staying in the middle lane. As usual the right hand lane was stalled with vehicles waiting for gasoline. Checking the pricing as he drove by service stations the prices astonished him.

"Hold Smokes," he said, "Up another dollar." Now all the pricing was by the pint. Veeby thought that it made the consumer feel better, but they were suffering.

He was all right. As a contractor working for the Federal Government he had a special green rationing card. He drove slowly passed the Green Card station. Only a few vehicles, most new and shiny, waited by the pumps. Neatly attired uniformed attendants rushed around wiping windshields, but all stayed away from the customers. Social distancing was ingrained. Pandemic after pandemic had taught people to avoid one another.

"Poor fools," Veeby thought smiling to himself. Working for the Feds had its perks, but still his contractor status lacked much to be desired. As he turned off Connecticut Avenue onto Lolliard Circle Drive and then turned onto Lolliard

Circle Court he saw his wife's Toyota parked in the street. It was blocking the driveway.

"Screw this," he cussed. He pulled up behind the Toyota and honked the horn.

Marion Veeby came to the front door and looked to the street. Veeby lowered the passenger side window and leaned over scowling at his once attractive wife.

"Pull forward!" he shouted, "Don't block the driveway!"

"Park on the street. There's a space out there."

"I don't want the car on the street. I told you!"

Vehicles left on the street were easy targets for roaming gangs of homeless kids. On any given night vehicles on a block would have wheels, radios, even windshields ripped out. Many times they stole the whole vehicle even with homeless people sleeping inside.

Homeowners had set up light beam security devices across their driveways. Anybody entering the property tripped the system. Alarms went off, lights flared and sooner or later, usually within a couple hours, over-worked cops showed up.

But sometimes, the kids never knew, the cops had taken to moving around changing houses. When a snare wire was tripped five pissed cops came charging out of the house with shotguns blazing.

Even lawyers had stopped screaming. They had been robbed, raped and beaten so often that nobody, not even lawyers or the Liberal Press, pissed and moaned about Civil Rights anymore.

"I'll get the keys."

Marion constantly irritated him these days. One of her constant complaints was that they needed a garage. If she parked the car up further between the houses the wheels sank in mud.

Veeby understood. Yes, it was true that the area between the houses was low and usually wet. Neighborhood tenements further up the hill on the other side of the alley were in disrepair. Raw sewage ran in ruts.

The toilets were undoubtedly leaking and sewage was flowing downhill to puddle between the houses. On hot summer days the odor was three-dimensional.

But there was no room for a garage and since they were renting Veeby would be tinker damned to spend money on it.

"Pull up between the houses," he shouted again as he got his car in gear. He watched impatiently as Marion walked to her used Toyota. She drove her car up to the end of the driveway.

As Marion gunned the engine the vehicle jumped forward and as usual she brought the vehicle to a stop up too close to the house.

Veeby drove his vehicle up into the driveway behind the Toyota. He got out and walked up behind the other car while Marion sat in the Toyota with the windows closed.

"Come on Marion," he called to her sensing that she was angry. "We're going to move. This is temporary."

She didn't budge. Veeby hesitated halfway to the front door and then turned around. Her car was muddy. Brown ooze had splattered up on the tires. Marion sat with the window up staring straight ahead with her hands up on the steering wheel as if she were driving somewhere.

"Come on Marion," Veeby called to her as he walked to the drive. He saw her car jammed up between the houses. The front wheels were settling in muddy grime.

"You gunned it Marion. You spun the wheels. Mud splatters."

She lowered her car window. "You're so damn condescending! I know mud splatters!"

"Now you can't open the door! You'll have to climb over the shifter, over the bucket seats, and climb out the passenger side door."

Veeby took a deep breath disgusted. "You're precious!" he yelled sarcastically. "You got yourself stuck again. Honestly, Marion. Aw hell. I give up!"

When he turned around and walked by the Toyota his right pants pocket moved within a few feet of the exhaust pipe. As

Veeby took a few steps, disgusted with his wife, the instrument went click, click, click.

"Now what!" Veeby said irritably. "You have a bomb in the…"

He stopped, confused, suddenly realizing that the sounds had come from his trouser pocket not the Toyota. He fished around in his pants and wrapped his fingers around the instrument.

"Marion!" he shouted reaching into his pocket and grabbing the device. He knelt down by the car while fumbling with it. Moving it slowly around the vehicle Veeby held instrument close to the metallic blue metal of the car body and then down by the exhaust pipe.

Close to the Toyota's exhaust the device clicked loud and fast. In his reaction, Veeby frantically pulled his hand away. He cringed back in sudden fright momentarily losing his balance. Going down to one knee he vaguely heard the passenger door slam. Marion climbed out staring at him.

"What now?"

"Get away from the car!"

"What?"

"Now!"

Veeby got to his feet and pulled her away. His eyes went from her to the house. He stood ten feet back from the Toyota, his jaw slack, with his left hand on Marion's arm. He stared at the instrument in his right hand. His mouth formed the words, 'My god.' He held the instrument close the automobile. His left hand went up in a gesture of helplessness.

Marion stood with her hands on her hips. She said something, but Veeby's stunned brain did not hear. She dropped car keys into his pants pocket

"You want that damn thing moved again," she said, "Then move it!"

When she walked around him he reached over taking her arm and forcibly turning her to face him.

"Donald so help me," Marion faced him, "If this gets physical I'll get my brother over here."

"Marion!"

She looked at her husband. His face was contorted, but not with anger. This was fear. She had never seen a look like that. His expression was suddenly that of a very frightened man. His gaze went from the Toyota to the chattering device in his right hand.

His eyes wide, his mouth moved, but no sounds came. He couldn't seem to take his eyes off that damn gadget that was making a clicking sound. She knew about his work. She knew the instrument measured radioactivity. She knew that the clicking...

"Donald?"

He held the instrument at the Toyota. The device chattered as radioactive particles hit the receiver and then as he swung the device away the clicking slowed and finally stopped only to resume when it pointed it at the exhaust. He circled around her and pulled her away from the vehicle. He got between her and the car.

"The car is hot!" he shouted.

The gauge needle was in the red range indicating at least two hundred rads of radioactive particles—lethal amounts of radiation were streaming off of the automobile.

Donald looked again at the instrument. His eyes were wide and unblinking. Then he quickly looked at the car and the house.

"Call 911. Hurry. Tell them we have a ..." She tried to pull away, but he suddenly held her close to him.

"Marion," he said calming down becoming the scientist again, "Please, tell me that you brought gasoline this morning. Yes? Was it this morning?"

"Yes Donald," she whispered hugging him, "It was this morning. We get gasoline once a month. Today was the day."

"At the Green Card station?"

"Yes Donald." She put her hands on her hips.

"Where else would I find gasoline?"

21

"Thank god for rationing!" Donald shouted.

Marion brought one tank of gasoline per month for their second car. Donald ran his fingers through her hair. Strands of her hair did not come off in his fingers. He stood with his back to the Toyota sobbing.

His gaze went from her to the Toyota to his own vehicle, wondering if maybe it was radioactive and realizing that it wasn't or he would have heard it.

His pesky neighbor Vern Anderson came out of his house. He strode across the front lawn with his arms waving. He was always complaining about something.

"Squabbling about parking again, eh?" he sniveled, "Well, we can hear you two all the way into the goddamn kitchen. I can't see why…"

Veeby walked away from his neighbor and looked hard at the house. The kid's bedroom was right on the side window facing the Toyota. He dropped the instrument to the ground. "Get the kids out!" Veeby commanded.

Marion ran for the house. Veeby brushed by Anderson. He yelled to stay clear of the Toyota as he ran to his own car, got in and drove it out into the street.

At seven: thirteen that night the phone rang on a night editor's desk at the Washington Post. Mac Phillips, a night editor, answered. It was a slow night. Phillips slouched at his desk.

"You should send a reporter out to Lolliard Circle Court in Chevy Chase," a man said. "It's right off Connecticut Avenue in Chevy Chase."

"What's your name Sir?"

"Anderson," the caller said, "Vern Anderson. Listen, I'm telling ya. They've got trucks and crews out here. I'm telling ya. It's a circus."

"The circus is there? Where? In Chevy Chase?"

"No, gall-darn it!" Anderson shouted, "The authorities. The words H-A-Z-M-A-T, he spelled it out, are written on all of the vehicles!"

Thirty minutes later Phillips and a staff reporter were on Lolliard Circle Way about a block from Lolliard Circle Court. Down the street trucks with massive searchlights illuminated people running in and out of bright puddles of light. They ran in and out of houses tossing merchandise on the lawns.

Women and children scuffled with police and other personnel in white radiation suits that were, apparently, trying to stop people from entering their houses.

A rabble of voices floated through the evening air. Phillips heard men shouting threats while women and children scuffled with squads of police that charged up onto lawns.

Police chased people in front doors and outside doors. People ran in and out yelling at the police and one another.

People dragged furniture outside and tossed it into their front yards. The women yelled high-pitched hurried instructions to children that yelled at one another, cursed the cops and followed the women into houses. The whole neighborhood seemed to be running around in the night. Panic was clear in their hurried actions and in the eerie waves of noises, glass shattering, household goods became debris the instant they hit the ground.

Loud speakers blared. Police shouted. Yelling and wailing and crying people sent waves of stressful sounds undulating through the night. Cops barked orders with a PA system as kids ran up throwing rocks at them.

Mac sat in stunned silence as workers in white radiation suits stumbled around and over furniture that had been tossed pell-mell into front yards. Piles of broken furniture lay scattered over lawns and sidewalks.

Four or five people, including children, had two of the white-clad workers down on the grass, stripping them of their white garments. In the glare of harsh police lights a little girl donned a huge headpiece, bolted into a house only to reappear a moment later with something in her hands.

"Is that a doll?" the reporter asked. Mac watched as workers ganged up on the girl fighting for control of the doll. The protective headset lay forgotten on the grass.

Another radiation headpiece got pulled off a worker's head. It plopped to the ground and was promptly trampled. A uniformed cop joined the scuffle, yelled at the combatants, tripped over the headpiece and kicked it into the middle of the street.

Fog, gray and silent came rolling down the street, slowly obscuring the scene. It flowed in front of the searchlights casting people and equipment as shadows darting here and there.

"Give me that camera in the back seat," Mac said to the reporter. "We need pictures of this before the fog settles in."

Getting out of the vehicle Mac only took a few steps toward the activity when a cop came out from behind a truck.

"Street closed!" the cop yelled. "Get out of here!"

"I'm with the Post," Mac said. He raised his camera as proof, but the cop rushed up snatching the camera.

"Street closed!" the cop yelled again. "Move it!"

As Mac watched, the cop turned around and threw his camera onto a lawn with other junk. The cop then ran around the front of a fire truck while Mac turned around and pulled out a second camera.

Four or five kids, fast moving blurs of speed angrily milled around the cop. The kids threw some rocks and then, apparently acting under some common unconscious impulse, ran away yelling incoherently.

Mac looked passed the fire truck. The same cop saw Mac again and approached him. Mac showed his pressed pass to the officer. "What's going on Officer?"

"I told you to move it!"

"Who is in charge?" Mac asked.

The cop stared at him and over at the reporter. Mac had the impression that this cop was a very angry man.

"Captain Hearn's is in charge. He'll have a briefing later."

"Can we get down there?"

"Turn around," the cop said. "Last warning!"

As the cop walked away and then ducked as a wooden chair came sailing through the air at his head. The chair splintered against the fire truck. The cop ran after the kid that had tossed the chair. Mac phoned the Post.

"Stan?" he said when the dispatcher at the Post answered, "I want a helicopter over Lolliard Circle Court in Chevy Chase in ten minutes." He gave directions to the dispatcher.

Mac leaned out the car window searching around in the fog for other signs of reporters. *Where were the news copters?* Why were they the only reporters on the scene? When he got back on the phone a man's voice broke into his conversation.

"All aircraft are grounded."

"By whose authority?"

"The President of the United States."

"What's going on here? We're the Press. We have a right to know!"

"Get out of here or go to jail," the unidentified man said.

Phillips flipped the phone closed. He saw that the cop had his ear next to a small radio on his shoulder. When the cop moved in his direction Phillips raised his hands in a meek gesture of compliance. He got the car in gear and quickly drove away.

*   *

One hour earlier at the White House the President sat in the Oval Office with five of his most trusted advisors. He yawned as he checked his watch.

"OK, I guess I'm still tired. Tell me again. What?"

"A citizen's vehicle was discovered to have radioactive gasoline. It's hot."

"What?" the President asked blankly, "The gasoline?"

"The vehicle Sir. The gasoline is the source of the problem. The vehicle, a Toyota as I understand it will have to be encased in lead and buried."

The President rubbed his face again. "Look I just got back from Israel. We have another crisis brewing over there. I've got jet lag. What? A radioactive Toyota?"

"The fuel was purchased at a Green Card station," Clarence Stanton, a senior advisor stressed his words. "As you know, Mr. President, government needs come first. Green Card stations get the fuel first. The fuel was purchased in Chevy Chase. We're tracking the supplier."

"You say the Toyota will have to be buried in lead?"

A murmur of agreement circulated around the room.

"And concrete," volunteered another.

The President put his head down, closing his eyes and for a while those with him thought he had dozed off. Moments passed and nobody spoke for fear of disturbing him.

"Who owns the car?" the President finally asked without opening his eyes.

"A contractor," Advisor Stanton answered, "A PhD chemist. His name is Veeby. Donald Veeby. Works for the Department of Energy."

"He a target of terrorists?"

"It's doubtful," Stanton said. "He's a low profile contract scientist. He doesn't even rate health insurance."

Stanton stopped speaking realizing that, given the situation, Dr. Veeby and his family might go to the media. He might cause trouble about the government's handling of the never ending War on Terror now in its thirty-third year.

Maybe, just maybe, giving Veeby some job benefits and a senior career position might be the best course of action.

"About this man," the President said. "He's a chemist with the Department of Energy?"

"Yes."

"We might want him on our side," Stanton explained.

The President shrugged. "The Department of Energy?" he asked incredulously. "What harm can he do?"

"He could make trouble. Start blabbing about the government's handling of the War on Terror. I mean given that this situation is shaping up to be a terrorist attack."

Others around the room seemed to be thinking the same thing. Another round of murmured agreement echoed to the

President's ears. His tired brain made the connection between the chemist and the possibility that he could make trouble.

"We want this guy on our side! Do what needs doing."

"Yes Mr. President."

"I hate whistle-blowers! They make trouble."

"Yes Mr. President."

"We need to control this!"

"Yes Mr. President."

"One car is involved?"

The assembled sat with heads bowed. Nobody spoke.

"Then you think the problem is widespread? Others have purchased hot gasoline?"

"Yes Mr. President."

"From more than one of our Green Card stations?"

"Yes Mr. President."

"More than one vehicle?"

Again the heads nodded agreement.

"Anything else?"

"The citizen's house is radioactive. It will have to be torn down and the rubble buried with the car."

For the first time the extent of the disaster seemed to penetrate the President's thinking. "Good Lord," he said. "Houses?" he asked plaintively. His eyes were wide open now and searched in vain for some good news. "Many houses?"

"That's what we've got to find out."

"We've traced the gasoline back to a station in Chevy Chase. We've set up a perimeter," said another.

"A perimeter?" asked the President.

"For two blocks around the installation. All the underground tanks are radioactive. The whole area around the station is deadly."

The President listened quietly and then rubbed his face again, "But, but, My God!" he stammered, "We can't bury the whole goddamn gas station! Er," he looked anxiously around the Oval Office, *"Can we?"*

In time the Americans traced the radioactive gasoline back to the Kamsi Kingdom. Kamsi Terrorists had dropped three

Russian nuclear bombs down oil wells. They had sabotaged one of the last three huge remaining oil reserves, known in the trade as 'elephants.'

The terrorists were identified individually as high-ranking personnel in the Kamsi Royal Corp of Engineers. In the end the Americans had been buying radioactive oil from the Kamsi for over two weeks.

All the oil had been 'cracked' into various petroleum products right in the Kingdom, because Americans didn't want smelly chemical refineries in their neighborhoods. Or their cities. Or their states.

In any other age or with any other product besides oil such an act of terrorism would have meant war, but now the weakened fragmented United Remnant, of what had once been the United States, put out an appeal for help.

Many countries did what they could, but most were in crisis with no jobs and dwindling energy resources. As living standards plummeted governments could no longer afford even basic services to citizens.

But then, astonishingly, the Kamsi Kingdom offered millions in gold dinars to help with reconstruction. The President didn't understand it, but surmised that some sort of power struggle had taken place in the Kingdom.

The most nagging problem was what to do with the ships, the oil tankers that had carried radioactive gasoline and other oil products to the Remnant?

Negotiators at the United Nations knew that the ships had to be sunk *someplace;* and yet, every other country in the world, including Russia and Mexico, threatened nuclear war against the Kamsi if those highly radioactive ships were sunk anywhere around *their* own territorial waters. Only the United Remnant begged for more oil. And that, as they say, was that.

With financial aid from the Kamsi Kingdom the Americans mounted radioactivity-monitoring equipment on aircraft and flew a grid pattern over most of what had once been the Eastern United States.

One hundred seventy-three thousand automobiles and trucks, buses, tractors and fifty seven thousand gasoline-powered lawn mowers were identified as radioactive.

That was a catastrophe, but remarkably, because of the metal and plastic shielding provided by the automobiles and the fact, that due to gasoline rationing very little driving was done, most people didn't suffer radiation sickness. Their vehicles and houses were another matter.

Eight thousand four hundred and seven houses were condemned as unsafe for habitation due to high levels of radioactive contamination.

"I am not sparing any expense," the President insisted to the press. "We'll move to British Columbia if we have to make people safe."

When the Canadian Prime Minister heard about the President's remark he replied, "Over my dead body." He ordered the Canadian Army to protect the border.

"Who do those white bastards think they are?" he loudly asked the Canadian Media, "Mexicans?"

Of course there was controversy. Not over the Premier's remark, but over whom in America should shoulder the blame for the gasoline fiasco.

The Kamsi had committed an act of terrorism, but were making an effort to make amends. The Kamsi leadership seemed to honestly regret the damage done. The world governments, especially the United Remnant, knew they had to tread carefully. Not only was the Remnant broke, but it was also weak so it couldn't make war on anybody.

Everybody knew that there were fanatical elements in Kamsi society that never forgave the Americans for their behavior when the American military made war with impunity.

Americans thought they understood terrorism, but they didn't. Not until they were weak and broke. Now, however, political rulers in the Remnant had to tread carefully, because Kamsi society was rich, but also politically unstable.

Everybody, including all of the educated citizens in the Kamsi Kingdom, knew it. The Middle East was a powder keg and consolidation under the Kamsi did little to soften the feelings of the average citizen against the West.

The money they gave to help repair the damage was considerable. The Kamsi Kingdom was, therefore, immune from condemnation and blame. No government on earth dared to openly condemn the Kamsi. They were above reproach. Otherwise, however, the blame for liability spread far and wide.

"Look," a leader of a street gang said as he gestured at all the automobiles parked on the street. "Ain't that a kicker? The dumb bastards are parking on the street again."

"Screw'em," said another. "I ain't touchin' it."

"What's ta be afraid of?" asked Boss. His smile showed green teeth set in red gums. He lifted an instrument out of his pocket.

"No shit," said one. "What is it?"

"Picks up something called radio something," Boss said.

"Radio…what?"

"Never mind," Boss said. "When this thing squawks scram. Got it?"

"Does it work?" asked another.

"Sure," said Boss, "I got it from a guy over on Lolliard Circle Court. He showed me how to use it."

3

As Chief Scientist in Agricultural Genetics at the University of Tel Aviv Jacob had helped negotiate the deal over Solbean with the Americans. He was also responsible for Solbean development, crop testing and production. Although he still did research much of his day was involved with management and politicians and with the media.

His was the face on the publicity about the wonders of Solbean. While he took great pains to praise the work of Herman Burger and Emil Zada and others it was primarily through his pioneering research that many of the scientific breakthroughs had been made; and, that is what made losing the crop so personally and professionally disastrous to him.

If Leviman or others in the ruling elite decided they needed a fall guy, a patsy, to throw to the wolves Jacob understood that he was made to order.

Even worse the public didn't know about the crisis, because the public had been told that the oil supplies from the El Kamsi sect in the Kingdom was reestablished and that the Solbean crop was a 'backup' source of fuel.

The public didn't know that, in an effort to bring pressure on Israel, the Kamsi surrogate suppliers had stopped shipping oil to them over a month ago.

The Select Committee on Agriculture, which Jacob chaired, had voted against telling the Israeli public that, as usual, the Arab population was at war with Israel. Jacob had fought against secrecy and in the process had made more enemies.

"Why not tell the public the whole truth?" he asked.

"Not advisable," Leviman had cautioned. "With no oil what if it fails? Solbean is an experimental crop. Yes?"

"Experiments fail all the time," Jacob explained. "We are used to hardship. With the truth the public will be prepared."

"You mean scientific experiments," Leviman said.

"Of course I mean scientific experiments."

"I mean socially. I mean the consequences of telling the people that an answer has been found, a cure-all for high energy prices, for shortages and for the high cost of food, and then..."

Leviman hadn't needed to finish his sentence. The Israeli public had been through too much to raise their hopes and then face disappoint again. Jacob stayed silent and by his silence he had agreed. That had been a year ago.

Now, with two good Solbean crops behind them the current crop was dying. No wonder Leviman had wanted to keep the whole thing secret, because secrecy would make sense only if Leviman knew all along that, one way or another, the crop would fail.

Over the last few months Jacob had come to believe that either the Americans or Leviman was working very hard to sabotage Israeli efforts to become more energy independent.

Research progressed. Crops planted, field after field, and then the first and then the second crops were harvested. Solbean was everything Jacob had predicted. Yet over the months Leviman's attitude toward him had become increasingly hostile.

"Good luck," Masal said as the jeep pulled up in front of the Agricultural Sciences Building. As Jacob climbed out of the jeep he saw two razor sharp silver orbs fading away in vast star fields of the Milky Way. Those silvery moving objects so obvious against the stars were the American satellites on patrol over Israeli skies. They would return in ninety minutes.

As the jeep drove away Jacob stood alone. His gaze went to an enclosed stairway beyond the gate that led to a raised secured walkway. The walkway extended between the security building and his laboratory.

Sometimes security forces had dogs patrolling the raised walkway, but now it all looked deserted. Jacob held his security card in his fingers. He figured the odds of getting through without tripping the system.

Once he entered his security pass into the electronic lock Leviman's security would trace his every move through security sensors in the university.

His presence at the gate would be known. If anybody in the security office were paying attention he would be arrested on the spot for violating the terms of his house arrest.

Jacob put the plastic card back in his pocket. He ran down a narrow side alley between buildings and into a small courtyard. Moving cautiously through the courtyard he climbed over a fence and ran along the side of the Physics Building and up a flight of stairs. From there he went up a ladder to the roof and climbed over to the Agricultural Sciences Building.

His laboratory was on the roof, but nobody ever seemed to question why, over the past two weeks, since Talya, Herman and Emil had disappeared, almost all of his work was done at night.

That a plant researcher, a plant geneticist, was doing most of his research *in the dark*, even though plants need sunlight to manufacture complex hydrocarbons, never seemed to cross the minds of the politicians that controlled the University and the money.

Jacob had worked alone for the past two weeks. For the last two weeks Jacob did all his own experiments. The advantage of being a political outcast was that it left him free to pursue his own interests: and his interests centered on those two Chinese built satellites that were the key to Israel's agricultural future.

Jacob had only one task tonight: Show the Americans that they could not hold Israel hostage to their control freak paranoia. By breaking the satellite lock on the Solbean crop Jacob felt he could break whatever political hold the Americans had on Leviman. After tonight the satellites *would* radiate the crop.

Jacob walked out onto the darkened roof and felt his way along and settled down on his knees in deep shadow, listening to the hum of a distant electrical generator.

He wondered how many more days that generator would have fuel if the entire Solbean crop failed. He looked skyward and waited and in about ninety minutes the two satellites rose over the eastern horizon.

Two bright silvery specks moved up from the horizon speeding westward and were almost over Tel Aviv…Tel Aviv that was always listening.

Jacob entered another rooftop structure, found a notebook by flashlight and walked back to the satellite arrays facing the night sky.

He entered control codes into a computer keyboard and satellite dishes swung on their mounts and stopped as their electronics found the two small answering targets in the sky. Jacob uploaded encrypted computer instructions, but before he could find out if his calculations had been correct pain radiated up his spine.

It never entered Jacob's thinking that others might be waiting for him in the dark. Jacob's knees buckled as hands of steel caught him as he fell forward. Darkness filled his vision. He lost time.

"What have you done?"

Jacob stood slumping between two men. He was barely conscious, but recognized Leviman's voice. Jacob opened his swollen eyes to see daylight streaming in through wide high windows of a dingy industrial building. Jacob's stunned brain suddenly made the connection, as a steel door slammed someplace behind him, that this place was in a prison.

"What have you done?" Leviman repeated.

Jacob got his eyes open enough to see the President give a quick nod and Jacob cried out as a fist ripped into the muscles in the small of his back.

Jacob pitched forward onto a makeshift desk. The President, sitting on the other side of the desk, leaned forward, taking Jacob's hair and lifting his face to him.

"We are going to find out."

"I'll tell you," Jacob blurted out.

Blood trickled from his swollen lip. At that instant his greatest pain was his tongue. He had bitten it when the man hit him.

"It makes no difference now."

"So?" Leviman said. "So tell."

A man tightened his grip on Jacob's head squeezing hard.

"We Israelis are good at finding things out," Jacob whispered. The man let go of him and Jacob put his face down to the wood. Leviman pushed his face away giving a curt command to his men. They dragged Jacob off the desk and held him up between them.

"So that's it," Leviman said. "You spent weeks on that damn roof measuring radiation coming off of the American satellites. You broke the American control code, their signal, to the crop?"

"Yes. Now we control. Not the Americans. They can't shut us down."

"You broke their code?" Leviman asked again.

"Yes."

"For what? We have the other variety."

"But we can't use it," Jacob whispered.

It took a moment for Leviman to figure it out. Jacob saw a slow realization flicker across the President's face.

"We must use the first variety to capture the market," Jacob whispered. It was only the variety needing microwave stimulus that poisoned the soil.

Leviman stood and leaned forward putting his hands down on the desk to see Jacob's face.

"You broke the code!"

"I did a damn sight more than that!" Jacob said finding new strength in his voice. "I made sure the Americans couldn't get control. Those satellites are ours! We control them!"

"And the Solbean?"

"It's safe!"

"Then why is it dying?"

Jacob forced his weak legs to hold him up. He shook loose of the men. One of them grabbed him, but on a command from Leviman he stepped away. Jacob stood on wobbly legs facing the President.

"It was the satellites!" Jacob insisted.

"No it wasn't," Leviman said. "We've monitored the satellites too. You think we're unconscious?"

"Control," Jacob insisted. "We need control."

Leviman gestured to his men. Jacob felt himself lifted up and then pushed over the desk with his face down hard to the surface. Leviman leaned down to him.

"Tell me," Leviman whispered to him. "Tell me everything. The crop is dying! It's not the Americans! You are the chief scientist! I want answers!"

Jacob tried to get his head around to face the President, but the guards held him firmly.

"I saved our crops from American control. Our crop is safe!"

"Then why is it dying?" Leviman asked again.

"I stole their system! It will take six months for the Americans to repair the satellites. I broke the code," Jacob insisted. "Solbean is the key to our survival and critical to the Americans. Solbean can get us away from the Kamsi and their stranglehold on the remaining oil. I had to act!"

Leviman grunted. When Jacob forced his head around to look at Leviman the President's eyes glistened. Yes, Jacob hurriedly thought, Leviman may have been working his side of the street. Is that what he would say to the Press? That Doctor Ebbtide had taken money from the Kamsi and destroyed the Solbean crop to serve the interests of his Kamsi masters?

"Our crop dying!" Jacob shouted. "I had to do something!"

"This has been your mission since Burger and Zada went away!" Leviman scolded. "Yes," he added, "This game, this tango you think we have been dancing. You are as much a politician as I am. This constant intrigue! Over who will control the crop! Ha!"

"I did what I had to do!" Jacob shouted as he got his head around to face Leviman again.

"The satellites radiate the crops! The Americans cannot turn off the switch. We control!"

"This thing with the satellites," Leviman asked,

"Is it to cover your sabotage? To give you deniability?"

"How could it?" Jacob asked. "With our control the crop is saved!"

"No it's not!" Leviman shouted. "Look at the fields!"

For an instant Jacob thought, gauging Leviman's expression, as he was reaching over the desk to shoot him. Instead Leviman slammed a fist to the desk.

"You have it backwards!" Leviman shouted.

"The Americans were never a problem. Now you have sabotaged the whole system!"

"How?' Jacob shouted. "I've given us control!"

"The Americans have a deal with the Chinese!" Leviman's shouts echoed off the walls.

"They know that we have sabotaged their satellites! You think they are unconscious? You think we are all stupid?"

"Who?" Jacob gasped.

"The Chinese!"

"I did what I had to do!" Jacob felt his strength ebbing. He tried to repeat the same words, but his head slumped forward onto Leviman's desk and a black inky darkness floated in front of his eyes.

When Jacob opened his eyes he saw bare whitewashed walls. They were as dirty as the bare earth floor. He sat in the middle of the room captive on a hard-backed cane chair. His head throbbed. A single stationary overhead light cast a dim puddle of gray light on the earthen floor.

Jacob raised his head as two people in black robes entered. Their heads and faces were covered in the same black cloth customary with Kamsi with small slits for the eyes, nose, and mouth. Black beards protruded around the mouth openings. Jacob watched the men impassively. His gaze warily looked for weapons or implements of torture.

"You awake," one of them said.

"Ohhh," Jacob complained as the men's voices caused his ears to ring. His head was ready to burst. He had never felt such a headache.

Behind the two men the door opened and a small man in a neat western tweed business suit entered. Jacob watched the other two men step aside and lower their heads respectfully.

"I am Sheik Halim Kasan," the man said. "I speak for Sheik Abu Abud." He said something in Arabic to one of the others who quickly ran out and slammed the door.

Sheik Kasan walked behind the chair. He stood there saying nothing. Jacob knew of Sheik Kasan. He was in charge of the Kamsi Space Program that had been created by and at one time belonged to the Americans. The Americans had traded away everything to keep oil imports flowing

If, sixty years before the American public had been told that the Government would sell their aging fleet of space planes and that the government would discontinue American ambitions in space, contracting with the Chinese to launch satellites the American people would have laughed.

Nor did the Kamsi want the old shuttles and planes for space flight. They used them in school playgrounds as equipment for the children to play on.

The shuttle's metallic shells had children's slides protruding out of the shuttles to the ground. The shuttles were for youngsters to enjoy. Now, if the humiliation with the space shuttles wasn't bad enough, a new rumor had surfaced that China was negotiating for Alaska to help the Remnant with its crushing domestic and foreign debt. Naturally, the Alaskans were busy setting up their own Republic and frantically building up their army. Rumors were flying and the Alaskans were unhappy.

"I am Halim Kasan," the Sheik repeated while still standing behind the chair. The door opened again. A robed man came back to the room one, with his hands spread wide in a gesture of helplessness and said something in Arabic.

"Ah? Only one chair?" Kasan asked. "In this whole building only one chair?"

Suddenly another man entered. Both guards came forward; tore the tapes binding Jacob's wrists to the chair and yanked him roughly off the seat. Kicking his feet out from under him and holding him by the arms they sat Jacob down hard on the earthen floor. They swung him around to face the chair. The Sheik casually walked around and sat down.

"Mr. Leviman contacted us yesterday," Kasan said.

"He seems to think you want to be here. You want to work for us? Or that maybe you are already working for us. He is not sure. Neither are we. Who are you working for? We both want to know."

"I don't work for you!" Jacob said. "Why would I work for you?"

"I don't understand this gift myself," the Sheik said.

"You come to us like manna from heaven. You are a gift from the President of Israel. Perhaps you will help me understand?"

"I don't know," Jacob said. "Maybe you can tell me if Leviman is working for you?"

"Leviman?" The Sheik seemed confused. "Are you playing a game?" At a quick nod from the Sheik one of the robed men came forward kicking Jacob hard in the back.

Jacob groaned. He rubbed his tailbone where the steel boot had bruised him. Straightening his back he looked up at the Sheik, because he knew that any sign of softness would bring more of the same treatment. What ploy could he used to negotiate?

"I'm an American citizen! You are kidnapping me!"

The Sheik shrugged. "Kidnap," he mouthed the word slowly shrugging again. "Such a harsh word." He said something in Arabic and again a guard went out of the room.

"No Doctor Ebbtide. We did not kidnap you. Your government sent you to us. Apparently they were reluctant to shoot you." He shrugged. "I don't know why. Perhaps you can tell us that too?"

"I don't know," Jacob said. He moved around trying to sooth his bruised backside. "I don't know why I am here. Leviman sent me? Is that right?"

"Most certainly. He seems to think that you are a traitor. That you hate Israel."

"I love my country!"

"What country would that be? You are an Israeli and yet you worked for the Americans. You used to work for us also. Remember?"

"Sure. I remember, but that was a long time ago. I was young then and could take the…" Jacob hesitated

"Heat?" the Sheik asked. "Is Israel not hot?" He smiled showing a gold tooth. "The climate in Israel has been getting hotter. The whole earth is hot. Once fertile farmland has turned to desert. Only the Solbean grows. It is a gift from God. No?"

Jacob said nothing. When Jacob didn't answer the Sheik asked, "You are here. You came by truck overnight from Israel. But I don't know what you want."

"I am a dual citizen of Israel and America. Since I am not welcome in Israel I wish to return to America."

"You want to go to America?"

"Yes," Jacob said. "That is what I want."

"I understand," Kasan answered, "but we need to talk with you about our agricultural program. Since you are here. Well," he added gesturing with his hands, "We cannot very well turn away such a nice gift since, I mean, President Leviman seems to have something in mind by sending you to us.

"We have great need of your expertise. Our entire crop is in jeopardy."

He leaned forward bending down to Jacob. "As of today the American satellites do not come over our skies. Something is wrong. Can you help us with that? The Americans seemed to have lost control of their satellites."

Jacob lowered his head. My God, he thought, that damn program actually worked? He had reprogrammed the satellites. The satellites would signal the crop to grow, and

40

would do so in spite of American instruction. But now, damn them, the Americans had responded by shifting the satellite's orbit. The crops in Israel and elsewhere would die. Jacob was astonished how fast the Americans had responded.

He had anticipated some response, but believed he could out-maneuver the Americans. Leviman's goons had, however, grabbed him before he could finish with the Americans. Now thanks to Leviman the crop was in ruin.

"The satellites do not radiate the crops," Kasan said. "Without the satellites the crops die."

"Don't know," Jacob lied. "Maybe the Chinese sabotaged the American satellites?"

"Or maybe the Israelis?"

"Maybe," Jacob admitted.

"Or maybe you?"

"Maybe the Chinese changed the orbits," Jacob said. "I don't know what to do about it."

"Strange," Kasan answered. "You are the one man I am sure could do something about it."

"No," Jacob said. "There's nothing I can do."

"I see." Kasan shrugged. "Maybe," he said uncertainly. "On another issue we have developed seed similar to Solbean."

"Seed?"

"Purchased from a source in Israel."

"Seed?" Jacob asked again. "You've gotten seed?"

"Yes, but it's inferior."

"Inferior seed?" Jacob asked. This was news. Jacob wondered if it was Leviman that had beat the Americans to a deal.

It troubled him, because Jacob doubted that even Leviman knew that another variety of Solbean was in testing. A variety of Solbean that didn't need radiation stimulus? Jacob knew that the Israelis had it, because he developed it. He doubted that the Kamsi had the scientific skill to develop such a plant.

"Have things gone wrong?" Jacob asked innocently.

"Things have been going wrong for a long time. We think it's the Americans, but something is wrong. Our Solbean thrives for a while and then dies. I think you could help us."

"Are you developing your own seed?" Jacob asked.

"The Israelis sold us seed."

"Israelis?"

"Through the Americans."

Jacob was baffled. That couldn't be true, because as far as he knew the Americans only had limited test seeds. Not enough to sell. So where were the Kamsi getting seed?

But, if it was true it was a direct violation of the agreement between Israel and the Remnant government. No wonder the Americans responded in haste. First the Israelis sell seed to the Arabs and then attack the satellites. The Americans must be wondering what in hell was going on.

To kill the crop all the Americans had to do was instruct the Chinese to switch off the satellite signals. Without energizing radiation the crop would die. But now the Americans probably believed the Israelis had sold the Arabs the type of Solbean that doesn't need the radiation stimulus.

Another part of the puzzle fell into place: If Leviman sold seed to the Kamsi he needed a sacrificial to cover his action. A way would have to be found for him to shift his guilt onto someone else when the Americans started asking pointed questions about where the Kamsi got the seeds.

The Sheik leaned forward and tapped Jacob on the shoulder. "Your efforts last night," the Sheik said, "Oh yes, we know about you and the American satellites.

"Leviman told us. The Americans work through you. You know their plans. You and they have made a long effort to sabotage them. And us."

"You?" Jacob asked.

"We got both types of seed," the Sheik said. "With the satellites gone the first type dies. We developed our own seed from the other Solbean, but now it is dying. We don't know why."

"You got the seed from Leviman?"

"Not from Leviman, but we had a long talk. Even Leviman agrees. We can't understand why you ruined those satellites."

"I didn't ruin them! They should provide the radiation stimulus the Solbean needs. I made it impossible for Americans to turn them off! I ruined nothing!" Jacob shouted.

"But then the Americans asked the Chinese to change the orbits of their satellites. They did. The Israeli crop is dying. Both varieties. We know. Leviman told us," the Sheik said. "Most of it is already dead. Leviman has evidence against you. This, of course, is good news for us, but we want to know the source of our good fortune. What is your plan? Do you have a plan? Maybe we can help you? So, if you are with us Doctor Ebbtide you must say. State your allegiance and your terms. If you don't work for us then what flag do you follow?"

"Israel," Jacob said. "Always Israel."

"Israel?" the Sheik asked. He hesitated a long moment apparently trying to understand the situation.

"I think I understand," Kasan finally said. "But now I want you to understand our side of things."

"I'll listen. Of course," Jacob said.

"First Israelis sell us bad seed. They take our gold and then laugh at us, but that thing with the satellites," Kasan said, gesturing to the ceiling, "It's all very clever. Even Leviman admits it."

"Leviman again," Jacob said.

"You did something to the American satellites. The Americans are very angry. They cannot reestablish control so the Chinese have changed the orbits. You did something, but why? Leviman wants to know. So do we. Why not explain what you did? Why did you attack the American satellites?"

Jacob sat on the cold floor staring up at the Sheik. When he said nothing Kasan straightened up putting his back into the chair, but held his gaze on Jacob.

"I made a mistake," Jacob said. "It's that simple. I thought the Americans were killing our crops."

"But you measured the satellite signals?"

"Yes."

"So you knew that the satellites were radiating the crops?"

Jacob put his head down. To his mind there had to be more happening; the Americans had to be behind the crop failure, because if it wasn't the satellites then what was it?

"There had to be something else," Jacob said.

"Some interference with the signals. Something causing the crop to die. Something. I don't know what."

"You are confused?"

"Yes," Jacob said. "I am confused."

"And so are we," Kasan said. "Israelis take our gold," the Sheik repeated, "Sell us bad seed and then turn their backs and make comments about those stupid Arabs!"

Jacob wondered what he was supposed to do about it. He sat at the feet of a Kamsi Sheik listening to him grouse about his troubles.

The same bastards that poisoned American gasoline, caused untold hardships and misery, had the gall to find fault with others?

Did this Kamsi bastard actually think he had a right to beef? Of course the Americans and most Israelis worked hard to prevent the Kamsi from getting Solbean. What the hell! Were these people crazy? If somebody screwed the Kamsi on a deal and sold them inferior seed so much the better. What did they expect? A red carpet? Jacob raised his hands.

"You expected something different?"

"Expected?" Kasan thought for a moment. "We spent millions of gold dinars. What did we expect?"

He let the question hang in the air and then brought his attention down to Jacob.

"We expected to work toward the future. We expect to solve many of our technical difficulties with the help of talented people."

Again Jacob sat passively watching the Sheik. What was *he* supposed to do about it?

"We know," the Sheik said, "that you have worked long and hard to deny good quality Solbean to the Kamsi!"

"I had nothing to do with that!" Jacob shouted. "I didn't even know that Leviman was selling seed."

"But it wasn't Leviman."

"Then who?"

"Israelis," the Sheik answered. When Jacob said nothing Kasan leaned forward. His voice harder, more incessant of a straight answer.

"You inventory your seed?"

The question caught Jacob off guard. Of course they inventoried the seed. Every research project succeeds or fails on the organization's ability to manage all of the elements—and certainly keeping an accurate inventory of seed varieties is essential to success.

"Did you really get the seed from Israel?"

The Sheik looked up at the light fixture. He blinked as if he were suddenly caught off guard. He seemed to realize that he was talking to the one man that would know or would have been told that seed was missing from inventory. Nobody had told Jacob about any missing seed.

"It is my understanding that the seed came from Israel," Kasan said at last.

Jacob had the feeling that he was truthful. The Sheik was a junior player in the high-stakes game of biofuel and doubted that he would have played any direct role in getting the seed or negotiating a deal.

"It wasn't Israeli seed," Jacob said. "I think I would know about that."

Jacob knew that whoever killed the crops, both Arab and Israeli, might be using those disasters to cover the theft of seed that was sold to the Kamsi. He was only one man. He couldn't check everything.

Jacob's rump was cold and sore from the kick and from his spot on the floor. It galled him to sit at someone's feet. Especially an Arab. He moved around uncomfortably, but at a command from the Sheik a guard touched him with a boot. A little kick that meant for him to stop squirming.

"I have no deal with the Americans," Jacob said

"I don't know why the Israeli Solbean is dying. I wasn't told about it for a week. I concentrated on the American satellites. I thought the Americans were ruining the crop with the satellites."

"Why," the Sheik asked. "Could you not tell that the satellites were radiating the crops?"

"Of course," Jacob said. Did this Arab think he was stupid? "Of course," he said again, "but the Americans are clever. They can change the phase and the frequency of the transmitted microwave signal."

"Could you not measure it?" Kasan asked.

"I did measure," Jacob said. "I found errors in the signal. Either the Americans or the Chinese were corrupting the signal. That's why I took action."

"But you took control of the satellites," the Sheik explained. "Yet the crops die."

"The crop was already dying."

"And?"

"I changed the frequency to the right signal," Jacob said. "And yes, I took control of the signal so that the Americans couldn't do anything about it. The signal was locked."

"You locked the signal?"

"Yes. That's right. They could turn it off and they couldn't alter the frequency. I captured the satellites."

"Except of course for the orbits."

"I didn't think of it."

"As you say. You are one man," the Sheik said.

Jacob knew that the Sheik expected good answers, but what could Jacob say? That he expected the Americans to come crawling to him? That he reprogrammed the satellites as a way to show the American or the Chinese that they couldn't blackmail Israel?

Jacob swallowed hard. He understood how it looked to Leviman and how it looked to this Arab. The satellites had been reprogrammed and everybody, Leviman and the Arabs, knew Jacob had the skill to do it.

It was timing. That's all. Just timing. If Leviman hadn't grabbed him in the greenhouse Jacob would have negotiated with the Americans. And, he would have negotiated from strength: If they changed the orbit the satellites would have failed completely. They would have been worthless.

But Leviman had grabbed him just as the damage was done, but before constructive action could be taken. It was timing. Events had outpaced him.

Now Jacob sat on the floor at the feet of an Arab. He regretted ever touching those damn satellites. His face reddened as Jacob thought of Leviman. The man was a gangster! How could a man like that become President of Israel?

It was outrageous that Leviman used gangster tactics on a scientist. Jacob sat with his head over feeling righteous anger welling up in his soul.

"I never meant any long term harm to the satellites," Jacob said at last. "I meant only to bring the Americans to their senses."

"I think I understand," the Sheik said. "You are not a criminal. You are not a destructive man. We understand. You had a plan and things went badly."

"If I could just talk to Leviman I could make him understand."

"He will not talk to you. Doctor, I tell you this in complete sincerity. He expects you to go back to Israel. He expects you to go back. That's what he expects."

"To be shot," Jacob said.

"I do not think he means to shoot you."

"I don't trust him," Jacob said. "I don't like the man."

Kasan crossed his legs and made himself comfortable. He seemed for a moment to be lost in his own thoughts.

"Maybe you are right," he said. "Leviman seems to think that you are already working for us."

"Leviman knows better," Jacob said. "He has his own plans."

"Possible," the Sheik said. "But I think he wants you to prove that you are loyal. He expects you to return. Face justice like a man."

"I can't do that either," Jacob said.

"I want to understand," Kasan said. "I want to make sense of it. We can offer you money. Work for us. Your reputation in Israel is ruined. Why not profit while you can? You are a Jew. Yes? Jews take money!

"You take money from the Americans. Why not from us? The world seems to think you already are working for us. You are a clever man. You take money from the Americans. Take money from us. Why not?"

"I'm not interested," Jacob insisted. "Something happened in Israel. I don't know what. Maybe Leviman is working for you?"

"Leviman?"

"He sent me to you. I am his gift. Remember?"

Kasan shrugged. "He sent you do us yes, but I think only as an alternative to shooting you. And," the Sheik added, "I think if you go back to Israel he will not shoot you. I have no doubt of it. But there are other ways of dealing with people."

"Yes," Jacob said. "There's always prison."

The Sheik stretched out his legs. His knee-high black leather boots reached out around Jacob's thighs corralling him.

"Everybody makes deals, eh?" the Sheik said picking up the thread of his logic. "Like the good old days, no? The Americans. Always good for a deal. Yes?"

Jacob cringed. *That* was *exactly* his view of the Americans. Always 'up' for a deal, but also wasn't that exactly what people thought of Jews—always angling for bargains?

"I won't work for you," Jacob at last. He felt confused again as the Sheik nodded and seemed resolved to the situation.

"Everybody makes deals," the Sheik said. "I spoke with the American President last night. You did something to their satellites. Leviman said the same thing. Israeli crops are dying.

Ours too. You seem to be at the center of it. Everybody is angry with you."

Kasan smiled. "The Americans are *very* angry. You broke your contract with them. They would be very happy to have you show up in America."

4

"Yes," Jacob said, "Things are against me."

"Are you loyal to your people?"

"Yes," Jacob said simply. "I am loyal. Always loyal to Israel."

"I expect nothing less," the Sheik said. "A man should be loyal to his people and to his country. Only fools like Americans think that they are citizens of the world," but he let his head fall over to one side, questioningly, his eyes on Jacob.

"But then, Sir, tell me why you will not go back to Israel? It is your country. Why not risk going back? Leviman expects you to be a loyal Jew. He expects you to return."

"I want to talk to him," Jacob said. "I need to understand what's going on. He damn near killed me last night and now he delivers me to you."

"President Leviman has made certain things clear," the Sheik said. "He will take you back, but will not take phone calls."

"I don't want to be here!" Jacob shouted.

A guard moved around behind him coming closer and then moved away at a word from the Sheik.

"Well then I will tell you," Kasan said, "We have a deal with the Americans. They want you. I do not know if they want you to work directly for them or they want you in their prison system, but they are willing to pay for you."

He smiled a mirthless smile, "but I don't think you will be very happy there either."

"My home is Israel," Jacob said.

"Your own President sent you to us. In fact, Ben Leviman wants you to go back over the border to Israel. I have already said so more than once."

"If he wanted to kill me why didn't he do it?"

"He has his reasons," Kasan said. "Maybe he would kill you and maybe he would not. We don't know. We have dealings with Leviman and other Israelis from time to time. We think he sent you to us with the expectation that you would simply go back."

"Why?"

"We don't know. Maybe to simply surrender to him."

"No reason to go back," Jacob said. "I'm under house arrest. My career is ruined. Maybe if Leviman changes his mind my life is over."

"We won't send you back," Kasan said.

"Why not?"

"Don't be stupid Sir. Leviman will not pay to get you, but the Americans will. We will follow the money." He laughed. "That's Jewish of us. No?

"We take a tip from the Americans. And the Jews! Yes? Who knows? We make money with the Americans and make some goodwill. Eh?"

Jacob gulped, nodded reluctantly and thought, yeah, goodwill and money. *That's* an unbeatable combination. The Kamsi had trouble with growing Solbean. They needed American help. They also needed Israeli seed. Good seed.

Jacob knew that sending him packing to the Americans was good business for the Kamsi, but might not be good for him or for Israel.

Without him at the University the development of new strains of Solbean will grind to a stop. Both Herman Burger and Emil Zada were missing and without at least one of them the program would collapse.

"Your situation is complicated," the Sheik said sympathetically. "Why not consider our proposal? We need your help. Why not consider it?"

For an instant Jacob nearly had a change of heart. What were his choices? He faced a crooked, but very powerful politician in Israel and an angry bunch of scientists and politicians in the Remnant.

He had, after all, put a huge monkey wrench into their satellites. Jacob thought for a long moment and saw the Sheik carefully watching him.

"So be it," the Sheik said. "Maybe there is too much history for us to work together. However, we will help you. We want your goodwill too."

"You will sell me to the Americans, but you want my goodwill?" Jacob asked incredulously.

"Can we not do both?" the Sheik asked. "We offer you opportunity to stay here. Work for us, but you say no. Would you rather we sent you back to Israel?"

Again Jacob shook his head. He needed time to think. He needed time to reason things though. He was a scientist. He needed facts, a clear set of facts to act on, but now all he had were enemies. Everything was complicated. Everywhere he turned he found people that made it clear they wanted him dead or in jail.

"It has been fifteen years since you were in America?" the Sheik asked.

"Yes. I left because..." he stopped, closed his eyes in sudden pain as he remembered the reasons for his sudden departure from America. The Sheik smiled again.

"Yes. We know. You had trouble with your wife."

"Private business! That's no business of yours!"

"Yes," the Sheik said, "but still fifteen years. Things have changed since you lived there. The situation there is much worse. The country has fallen apart."

"No thanks to oil," Jacob said.

"America," the Sheik said. "Of course now, after the Breakup, they call themselves the United Remnant. Without our oil life is hard."

Life *was* hard. Jacob was born in Boston and attended Harvard as a young student. Along the way he met many people especially police and lawyers and others on the 'other side of the law.'

Jacob cringed thinking of what a smart ass he'd been. In his younger years life was a lark and as far as he was concerned the booze would flow and the laughter would go on forever.

He wondered how much this Arab knew of his younger years, but dropped his gaze as he saw an inner light in the Sheik's eyes: His direct eye contact was a signal that one man gives another when he knows another man's little secrets.

"The law is waiting for you," Kasan said. "I am being honest with you," and tilted his head to one side.

"I believe you," Jacob said softly.

His memories drifted back to America, back when America was still what she used to be, but was fast becoming something that nobody wanted.

He remembered the apartment building in Old Boston. He remembered the bastard security consultant Mister M.C. Handy, hired by building management, to keep riffraff at least fifty feet away from the premises.

Meehan C. Handy scratched his two-day old whiskers. His eyes were deep hollow sockets set back in cheap black plastic glasses. He held up binoculars and trained them on a ragged lone man walking slowly on a rain soaked street below. Handy spoke mechanically into a portable phone, "Undesirable."

Jacob looked over Handy's shoulder. His job as a junior security consultant at Handy's company was to observe how things were done and to make design improvements to the equipment that Handy used.

Handy followed the vagrant briefly and then raised the binoculars upward toward a building roof across the street. On the other roof another officer, Ronald Puller, also had the vagrant in his sights.

"So it would seem," Puller said.

"Well then?"

"Okay," Puller replied.

In the apartment Handy put away the binoculars and picked up another telescopic piece of equipment. When he got it in front of his eyes and scanned the air between the buildings he saw a thin ruby line, a bright line of light, angled down from the building roof.

The laser light from the roof 'zapped' the man who slumped first to his knees and then struggled to his feet and raised a fist in Puller's direction all the while making a hasty retreat.

"You don't have many choices left," Kasan said.

"I think you're right," Jacob said. "I was thinking of my apartment in Old Boston. I almost took a job there when I was a student at Harvard. It's been years now."

"Would you do things differently?"

"Differently?" Jacob asked. Do things differently? Can we ever do things differently? I regret things. Sure I would do things differently."

Sitting on the floor at the Sheik's feet Jacob's mind was on overload; and, yet he also felt regret and anger with himself, because in his youth he was truly a better man: Resolute, confident. At twenty-three he would never let a situation like this throw him. Sure. He had regrets.

He remembered the vagrant. Even for a brief moment years ago, as a young student at Harvard, he had regrets about helping to design that damn laser security system.

Lasers are a wonderful invention. The light from two lasers can create a hologram, a three-dimensional image of anything. Lasers can even burn a unique number into the forehead of a bum trying to crash a building to find a place to sleep.

The laser was weak. All the tests showed that it couldn't actually hurt anyone. When it touched the skin, usually the forehead, it made a pattern of tiny holes, a million times smaller than a medical needle. That was the key to why it was harmless—the tiny laser holes were too small for the nerve tissue to react.

With security markings the movements of vagrants could be monitored with complete accuracy. Each number was unique. Each bum was positively identified. If a crime was committed, the police knew who to go after, because infrared video equipment could read the brands even through materials such as ski masks. The equipment to read a brand was truly remarkable.

The brand was three-dimensional and could be read from any angle. If the bum had his head up or down. With a hat or not hat. With a mask or no mask. The bum's own body heat gave the brand a 'presentation' that could be read anywhere.

Each pattern was a unique coded pattern, a number assigned by the national security computer network. Common names? Social security numbers? Unreliable! If vagrants were congregating in any large numbers that foretold a riot or other trouble the security apparatus could read and print out their names. There was no escaping the system.

Trouble was averted before it began. The laser 'marking' system was crime control and crowd control all in one neat package. Social activists called it 'branding,' but that was a libel. It was scientifically proven that the laser holes were far too tiny to actually cause pain when the brand was applied.

Oh sure, lawyers screamed about the system for a while, but it seemed as if the whole of Boston was filled with homeless people. After a while all the fuss about the brands died away.

Some vagrants had complained that they could feel the brand going in, but scientific evidence had disproved all such claims. The brands were a thousand times thinner than human hair and rarely, if ever, actually penetrated the skin. Nor would the skin grow back to obscure the markings, because the laser seared deep enough into the skin to block any growth of new skin tissue. It was a foolproof system.

Regrets? Jacob's heart sank as he thought about it. God! He had helped to create a system from which no person could hide his face. Every street light—those that still operated—contained a laser hologram camera. Mainframe computers worked day and night to track homeless people moving around the cities.

"Ah," Handy chuckled, monitoring the movement of the bum on the street, "That'll teach him."

Handy watched the bum and the others milling around outside the safety perimeter: Down on the street the vagrant shook his fist up at the building, but kept walking away. Handy followed him with the binoculars. "That fella is making a habit of hanging around here," he said.

He put down the binoculars, checked his watch, and looked out the window. Low winter clouds were coming in over the City and already it was getting dark. "Damn," Handy mumbled to Puller, "I've got to get to the store."

"Go now," Puller said. "I'm off duty in twenty minutes."

"Ah hell."

The neighborhood was badly deteriorating and more vagrants were found in the stairwells and sometimes in the hallways themselves, and nobody understood how they were getting in. But now Handy had to concentrate on getting groceries at the shopping center two blocks over.

"I'm going to the store."

"I'm going upstairs," Jacob said. "Good night."

Handy found his wallet. The door slammed behind that young fellow with the funny name. Ebbtide? What kind of a name was that?

In his wallet Handy found his plastic electronic security card for getting back into the building. "I got my card. So don't get trigger happy, hear me?"

Handy didn't hear a reply from Puller, but he left the apartment anyway. Once down at street level Handy stood inside the building's security doors.

Peeking out through dirty glass, he was worried. Evening was coming and with the loss of daylight anyone moving on the street was a criminal suspect or a target of criminals or gangs.

Still the refrigerator was empty. The day had dragged by before he knew it. Handy pulled the flaps of his overcoat up over his ears and stepped out onto the sidewalk. People milled about on the street, but it was cold and they scurried along as if they actually had someplace to go.

Twenty minutes later Handy left the store with his bag of groceries. His tired brain was stunned to see it was already dark. He tightened his coat again and cradled his bag of groceries in the crux of his arm. Rubbing his hands together, he braced himself against the cold. Down the block he stopped and stared in one direction and then another.

Handy was passing an abandoned building when a hard object caught him by the shoulder. Handy felt nothing except his legs melting out from under him.

Cold, terrible cold, froze his hands, but still by gently touching his mouth he could feel a swollen mass of pain where his face used to be. With great effort Handy stumbled to his feet.

His wallet was gone. His great coat was gone. Ditto the groceries. Handy had never known that cold could hurt so much. Cold bit into his flesh like thousands of knives. Using the wall of a building he struggled to get oriented. He got his bearing. He was only a block from home.

He got to a grassy strip in the middle of the road that separated his secure apartment building from the abandoned buildings across the street.

He made a dash to safety when he felt a dozen bees attacking his cold flesh. Even through the numbing cold the buzzing laser attack hit like branding irons.

"Puller!" Handy yelled raising his hands ineffectively as his forehead swelled up, puffing out in dozens of little bumps. His forehead burned like fire. Forcing his frozen fingers into a fist he shouted at the roof,

"Damn you Puller. I'll have your hide for this!"

"Wasting your time," a man spoke from someplace near him in the darkness. Handy spun around in a panic getting dizzy in the process.

"You live around here?" the man asked.

Handy felt panic grab his heart.

"Across the street."

"Stuck in your own mess?"

Handy moved around to get a streetlight at his back and take a good look at this fellow. There was something about the way the City engineers designed the streetlights: large puddles of bright white light were separated by absolutely black spaces. People stumbled even on smooth sidewalks. Handy stood on the grass and stiffened his legs.

"Listen," he said, "I've got to get across the street. I'm hurt. Been robbed."

"Yeah," the man said. "I see that."

Handy stood helplessly as man walked off in the direction of the grocery store. "I'm trapped too," the man said, "But I'll call security from the store."

"Hurry up," Handy yelled. "I'm freezing out here!"

"Don't try crossing the street! You're in the system now!"

"God damnit!" Handy cursed loudly. "I'm a consultant!"

"You know it and I know it, but *they* don't know it!"

"I'm a rich man," Handy said running forward into the street. "They can't…"

"Better get out of the street," another man said.

The voice was very close to him. Handy looked down to the grass to see a man wrapped tightly in a blanket. Another two feet over and Handy would have walked on him. The bum's deep resonating voice made Handy nervous, because the bum sounded intelligent.

"Step back," the bum said, "Or they'll give ya another brand to wear."

"Idiots!"

Handy looked around to see the other fellow walking quickly away toward the distant lights of the grocery store. Wind shrieked between massive brick buildings. Now, over the wind, Handy heard the sounds of wailing of police sirens in the night.

"Cattle drive," the man on the ground said. "Damn! I was getting warm."

Suddenly around Handy there were other indefinite shapes moving about in the night, standing, groping for their possessions.

Handy cringed as they brushed against him in the dark. When they went passed him they all said, "Cattle drive. Come on man. Git out."

Handy turned toward his building that seemed to beckon to him. It was just across the street.

"I can make it," he announced, "I'll run," but as he spoke he realized that the others were not listening.

Around him the shadows people moved away with astonishing speed and stealth. Handy saw see them only as shadows against a glare of his apartment building's harsh security lighting and shadows in and out of street lighting.

Handy spun around to face his building, put his head down and ran, but only made it to the apartment sidewalk when the headlights of two police cars blinded him.

"Move old-timer," a cop said. Handy felt a rough knob of a nightstick poke his back.

"Home turf Doggie," the cop said. He gave Handy another shove. "Stay away from that building. It's secure. See?"

"Listen you!" Handy shouted, but the cop, a big bruising menace in the glare of the vehicle's headlights, lifted his nightstick.

Handy jumped back as the motion of the nightstick blurred in his vision. Another cop came up. He was grinning.

"You know," the second cop said, "These old timers remember the good 'oil' days."

"Yeah," the other cop said, moving closer, his nightstick up again, ready to swing, herding a few stragglers including Handy out onto the grass strip in the middle of the street. Handy turned away from the cops and lifted his feet in a desperate struggle to make time ahead of the nightsticks

"Cattle drives," the first cop said. "This is small potatoes. Some cities have real cattle drives. I mean *real* cattle drives. In New York, hell," he said, "They used to drive twenty thousand head a night."

"I called it in," the kid said the following morning after Handy had spent the worst night in his life sleeping in an alley.

Later Handy confirmed that Ebbtide told the truth, but Puller, that bastard, had lingered on that damn roof until Handy had come back from the store. Now Puller was gone. Handy was left with Puller's gift.

"If I ever find that son of a bitch," Handy told Ebbtide. "I'll kill'em."

"One thing's for sure," Ebbtide said. He came up close looking at the brands. He made a face as he examined the burns.

"What's that?"

"Maybe you can't find Puller, but he sure can find you."

5

"Yes I have regrets," Jacob said. "I helped in a small way to develop those systems. Those brands! Oh god!"

"You try to make amends," the Sheik said.

"What?"

"Your work with the crop is commendable, but you want to make amends for past mistakes."

Jacob nodded. "I want to think so."

Sheik Kasan stared at Jacob.

"You see, Doctor Ebbtide that is what America has become. Every home, every apartment, is a reverse prison. Not to keep people inside, but to keep the world out."

"I doubt it's that bad," Jacob said.

"We really want your help. We know what you did to the satellites. Yes. We have our sources. Yes. We have sources independent of Leviman. We want the Solbean! Yes. It's competes with our oil. Yes. We want it for ourselves. That's why we need you. We need you," Kasan repeated. "We will pay you very highly," he raised his hands again, "We are prepared to negotiate," and he hesitated as Jacob studied his bearded face, "And yes we have regrets too."

"The gasoline?"

"Very much," the Sheik said sincerely. "It was a terrible act. Cowardly. We gave the Americans gold to make amends."

"I know," Jacob said. "I doubt it made up for the tragedy caused. They're still suffering from it."

"We can help each other Doctor. We have money. You have talent. The world needs better doesn't it? What do you say Doctor? You help us?"

Jacob felt a softening of attitude toward the Sheik and was now certain of the Sheik's sincerity, but it galled Jacob that if it hadn't been for the criminality of Ben Leviman he would still be at Tel Aviv University doing research.

Jacob shook his head. "I am loyal to Israel," Jacob said and he didn't say it, but he was determined not to be a 'bought bastard' like Leviman.

"I consider myself an honorable man," Jacob said

"I believe you." Kasan said "Yes Doctor. You are an honest man. We wouldn't waste time with you if we thought differently. You are good for your word." He gestured around the barren room.

"I apologize for this. We wanted a chance to speak with you. When Leviman communicated with us and said he wanted to send you to us we couldn't refuse could we?"

"No. I guess not," Jacob admitted.

"Without distractions. I wanted to speak with you face-to-face to clarify our situation to you. I have made my offer. The money does not interest you. You cannot be bought. You are, as you say, an honest man." Kasan waved a hand around the room. "I apologize for this," he said again.

"Apology accepted," Jacob said.

Jacob put his hands on the floor making ready to get up, but looked over again to make sure the guards didn't move, but decided to wait.

"Please answer one thing for me. One question before you go to America. Yes?" Kasan shook his head, "One question?" he asked again.

When Jacob didn't answer, he inquired, "Your divorce? That was under American law? Yes?"

"Yes," Jacob answered. "It was fifteen years ago," although it seemed very recent in Jacob's memory.

"Did you know that your wife is now a judge in Boston?"

"A judge?"

The news came as a rude awakening. His ex had been a mob lawyer. Gladys was a rough woman. Jacob knew he'd have to arrange his entrance into the country to avoid her. He knew from others that the best way into the United Remnant was through Germany, but they only provided flights into New Boston. Timing was everything. He'd have to move fast.

*   *

Jacob remembered sleeping in a jail cell in Boston, Massachusetts the day after he and Gladys had decided to divorce. Jacob had been arrested for dealing dope. Jacob was asleep on the cot when a uniformed cop came by the bars staring in at him.

"You have a visitor. Your lawyer is here."

As the cop moved aside, swinging open the door, Gladys his wife who was also his lawyer at the time walked into the cell.

"Ooh Jacob," she smiled, "I'm so glad I caught you home." She laughed at her own joke and dropped her briefcase on his stomach.

"Excuse me if I don't get up Gladys."

Jacob made an exaggerated gesture of dumping her briefcase on the floor. The snap broke and papers spilled out. A cop standing outside the cell grinned through the bars.

"You've done it again Jacob," Gladys admonished him. Jacob swung his legs off the cot.

"Gladys," he sat watching her as she knelt down for her papers. "I'm telling you again. I don't know where that bag of cocaine came from! I don't sell drugs! You know that! Gladys listen to me! You know me!"

She smiled at him. "Isn't that how it always is?" she asked playfully, "You think you know someone and then wham! Drugs are found!"

Jacob took a deep breath. "Those drugs were planted. I think you know that."

He watched as she shifted her face away from him. Her green eyes lost color in the dim light. Her fair skin drained of blood had a gray pall and her hair was more mousy than usual. He never did understand why she didn't use color, but then as a criminal lawyer maybe looking rough was a professional asset. Gladys scooped up the papers.

"I don't get it. Really I don't," he said. "I'll give you a divorce. Hell, Gladys I must have been dopey to marry you in the first place! Yes! I'll give you a divorce!"

He picked up some of the shattered documents and tossed them to the cot. "When can I get out of here?"

"I didn't put you here. You did. The D.A. has two witnesses that will swear you were selling drugs to them!"

Yeah, Jacob thought, two witnesses and both on trial themselves for dealing drugs. Worse, both were probably working for the American intelligence services. People in the FBI must be laughing.

Gladys represented both of the other two 'witnesses,' but Jacob strongly suspected that neither of them would serve any time.

She stared down at him. "Honestly," she huffed indignantly, "You don't think I would do this to get custody of Talya do you?"

"Yes Gladys," Jacob said, "Yes I do. You're a lawyer. You know the system! Those two bastards, those so-called witnesses, are working for you!" He returned her hostile stare.

"Can you prove it?"

"No."

"Then keep your mouth shut! Understand? In court, Jacob, do not make accusations. Prisons are filled to overflowing. Make nice. Understand? Be contrite. Throw yourself on the mercy of the Court. You have a very good chance of walking out. Nobody wants you in jail. Not me and not the Court!"

"But you do want our daughter. Don't you?"

"Of course!"

"Gladys! I work hard. I pride myself on being an honest man. Please! I need to hear it from you. I don't use drugs! Tell me that you believe me!"

Gladys shoved papers back in her leather briefcase. Grabbing the bag by the handles with both hands, she straightened up looking at him.

"I don't know what to believe Jacob. I married a young promising scientist. You won awards for your research, but then," her voice faltered. "Now, I find out about your other life. Your lies, your deception, giving information and secrets to the bastard State of Israel."

"But I've admitted to that!" Jacob protested. "Gladys I admit to that. Everything that I handed over, every detail, is known. I've done no harm to…"

"That not what the Government says!"

"Then why don't they prosecute me for espionage?"

Gladys stepped back holding his gaze. Jacob gulped in sudden understanding. The American Government, for reasons of its own, did not want to prosecute him for espionage; and, here he was in a jail cell. For a guy that loved to play games the authorities were letting him have it.

"Oh," Jacob said in sudden understanding.

"The government is playing its own game. They have you on their team."

"Honestly Jacob. I'm not on anybody's team, but my own."

Jacob felt certain that Gladys had contacted them. They had common goals: Gladys wanted Talya and the Feds wanted his ass in a sling.

"Oh boy," Jacob whimpered. He stood facing away from Gladys, because he was ashamed as she came up behind him.

"You are tainted goods Sir," she said. "In sickness and in health. For richer or poorer…" she said reciting part of their civil wedding vows. "I thought I knew you. Hell, Jacob, I can't tell anymore. Are you a scientist or a con man?

"The world is falling apart," she said. "Nobody, not even the rich, can afford to buy a broken car or a broken man. I'm sorry," she added. "Truly. Your involvement with the Israelis, with the Mossad, and my God, yes, I knew about your involvement with the gangs, but it's the espionage that has cost me everything."

Jacob wiped tears while keeping his back to her.

"I did what I did for Israel," Jacob said tearfully.

"Try to understand."

"Hell," she swore. "Sign the papers!"

Gladys left papers for him to sign on the cot and walked out. Jacob slumped down on the cot. Close by him the cell door slammed shut.

*   *

"Well?" Kasan asked. "Will you answer one question about your divorce?"

"You know my situation."

"Yes. We know," the Sheik said. "The Israelis are not the only people with an intelligence service."

"Then what?" Jacob asked.

"You wife is waiting for you. Our agents make that very clear. You left with your daughter. Your wife got you bail, but you missed your court appointment. You ran to Israel with your daughter."

"The whole thing was a frame. A setup."

"Eh? What?"

"A setup," Jacob explained. Kasan nodded, but Jacob still didn't think the Sheik understood why Jacob had to get out of America.

When he had been caught spying for Israel Jacob knew that any legitimate legal action against him might involve disclosing some of the material he had 'pinched' for Israel.

The American government played a double game. On one hand they didn't want him to disclose any of the scientific secrets he'd passed to his Israeli handlers. The Americans wanted to 'contain' the damage. But, on the other hand, they wanted his ass hung from a pole.

They wanted things both ways. No talkie-talkie, but lots of nailing ass to a door. And, right when they needed her in walked Gladys.

In his case the American government had apparently ponied up dummy drug charges. Maybe in cohorts with Gladys. The American government did not want him in court on charges of espionage. Israel is a friendly country and the only friend America has in the Middle East. Maybe the Israeli ambassador put in a good word for him? So charges on espionage were out, but drugs were another matter entirely. To Jacob's way of thinking everybody got what he or she wanted.

So the government decided not to prosecute him for espionage, but Jacob knew he'd have had to be an ironclad idiot to hang around for the second act. Between the government and Gladys his cup overflowed with vinegar.

The whole mess between the Israelis and with Gladys added up to more regret. Mistakes made and now in his middle years Jacob couldn't really figure out when the mistakes had started.

Everything, events and decisions, seemed to flow along with a life of their own. There were times when Jacob felt carried along on life's river like a leaf in the street after a heavy rain with the leaf heading for a storm drain.

"Look," Jacob said, "That was fifteen years ago. That's a long time. You're saying that my ex-wife is still mad?"

"She is angry," the Sheik said. "How to you say it, 'Mad as hell?'"

Sheik Kasan squared his shoulders, clasped his hands in front of him, leaning forward. "We will treat you well Doctor Ebbtide. We need you. We are prepared to sign a contract, good in any international court.

"You will be paid well. And," he pointed a finger at Jacob, "We will forgive the past and move on to the future. Americans and Jews are always making deals yes? What say you? We make a deal?" The Sheik stood up and in a gesture of goodwill offered the chair to Jacob.

Jacob's rump was sore from sitting on the hard floor. He nodded, got to his feet, and gratefully sat down in the chair again. It felt comfortable. The Sheik walked over and stood with the guards.

Jacob sat. His tender rump felt wonderful. His spirit seemed to get refreshed with his behind. Suddenly Jacob felt good and let his mind wander back to younger years and in many ways happier times. The whole world seemed to stretch out before him with a promise of wealth and unending good fortune. He discovered women when he was young and enjoyed every moment of it.

He had met Jennifer O'Hannon at a party at the Israeli Embassy. They dated. Jacob almost married her before finding out what her family did for a living. Her father, Frank, was in the business of crime and had connections throughout all the major intelligence services including the Mossad.

When some agency needed freelance help to give the service plausible deniability they could turn to Frank or any number of other outfits. But the Frank O'Hannon organization was considered to be among the best. They always produced results even if, occasionally, blood splattered wide of the mark.

Frank's daughter Jennifer was, at the time, Jacob's love interest. Jacob had wondered why Jennifer, who was quite attractive and almost thirty-five, had never married. She seemed to have men in her life, but her affairs didn't seem to last. Jacob wondered why. Then, just in time, Jacob had come into information about Jennifer's other loves.

Jacob was in the Harvard library one day when Ozzie, Jacob's handler with the Israeli Intelligence Services, approached him with some photos.

"We do work with Frank O'Hannon," Ozzie said. "You're dating his daughter. I give these to you as a friend."

"Thanks Ozzie," Jacob replied taking the brown envelope. Ozzie walked away before Jacob looked at the pictures. In with the photos was a brief note.

*   *

Frank O'Hannon stood leaning against the doorjamb. He watched his daughter hold the wood plaque with the human head mounted on it up high examining her latest wedding gift. Her eyes were alight with pleasure.

The head of her ex-beau was mounted on a handsomely polished walnut wood mount. Mathew's eyes were open. His gaze held that same stupid expression. He looked peaceful. Jennifer held the head up to the light and laughed out loud.

"Perfect! Have you ever seen this bum look better?"

"Baby," Frank said, "You're a class act. A real sport. How many women could get a head in a box and laugh?"

"Grabbing people for money is our game Daddy. It's a family business right? So why not get some additional benefit out of it?" She smiled up at Frank.

"Yeah, but Baby," Frank protested, "We use guns. You know. We're traditional."

Jennifer laughed out loud. "Daddy," she giggled, "You're so not with it! I don't use guns to get these saps where I want them. I have something better."

She lifted her head, posing, stretching a shapely leg out in the manner of a ballerina. Frank lifted the mounted head out of her hands, sat it down on the windowsill and looked at his daughter.

"Listen to your old man Jenny. You are playing a very dangerous game. I mean it. Any one of them could be a cop or have a gun."

"But you always come to my rescue." Jennifer pulled his massive head down and gave him a big hug.

"Baby, you're the best. You get them drugged. You bring them to the warehouse. I mean the con is beautiful. They think they are going to a chapel to get married and wham! You lead them inside and we're waiting for them, but Baby," he broke free of her, looking at his daughter, "We can get money out of them the old-fashion way."

He pulled a revolver out of his belt. "Huh?" Sliding the weapon back into his waistband he asked, "Why drag some jackass like Matthew to a fake chapel?"

"Hidden assets," she answered. "They tell me everything, Daddy. Admit it! We got more from Matthew and Peter than we ever would have." She winked mischievously.

"Admit it Daddy." She reached forward tickling her old man. He grinned and took her hands in his powerful paws.

"Yeah, well, I work with the boys you know. We watch each other's back. But you, Jenny, you work alone. Like I say you play a dangerous game."

"You're there for me. I know I can count on you."

"Yeah yeah, but listen," Frank said pointing down into the box, "We mounted his hands to see? Look down in there." Jennifer got to the box and found a matching set of mounted hands. She grinned her approval.

"A matching set? Ah, Daddy. How sweet of you!"

"Wasn't hard," he said. "Bobo did most of the work hackin' and sawin' and such."

"Gee Daddy you went through soooo much trouble." She kissed her father affectionately.

"Aw now," he said wiping away her kiss, "You know Baby, this sort of thing is like, you know, a perk. I mean, you know, of the business we're in."

"I'm proud of you," Jennifer said. "Most brides-to-be-have to be contented with candy and such, but you," she squeezed his hand, "Ah, Daddy, you're the real McCoy!"

"Now," Frank said turning serious, "You've got to stash this little trophy of your last wimp where your new wimp can't see it. So you probably don't want to keep it in here."

"Of course I'll keep it in here," Jennifer protested. "Hell Daddy. Jacob won't sleep here."

He hesitated and then asked, "Your new John got money?"

"His family is loaded. Owns a stake in a major motion picture studio. He's Jewish."

"Jewish?" Frank gagged on the word.

"Oh now don't fret," Jennifer reassured him. "Am I not part of the 'ol family team? I find'em, clean out their bank accounts, and," she went to the window, holding up the head, "Jacob will get his own headboard like the others."

She held up the headboard with one hand and patted Matthew's matted hair. Setting it down on a windowsill, Jennifer looked up.

"Jacob will get his own headboard. I promise, Daddy."

"That's my Baby," Frank said.

"Don't worry Daddy," Jennifer purred. "We'll take Jacob the same as all the others."

"That's my Baby," Frank said again.

Jacob Ebbtide

$*$ $*$

Jacob knew from grade school that he wanted to work for the Israelis. He had three reasons for this. First, he was Jewish. Second, industrial espionage is big business and that meant that there was money to be made; and, third, it would be fun.

'Joes' Beer Joint' was the last honest beer joint in Boston. No tables, only high chairs shoved up against plain walls, and the joint served nothing but cheap beer. If a patron wanted a sandwich they were told to drop dead. Ozzie nursed his beer.

"No food here?"

"Maybe some pretzels."

"Yeah, but I mean…"

"You always eat Kosher?" Jacob asked him.

"I'd rather eat pig than drink this swill." Ozzie looked around at the drab interior and the low class clientele.

"Why pick this place?"

"It's off the beaten path."

"It's off the beaten planet," Ozzie said.

He sipped his brew. Jacob took a long draught and wiped his mouth. "I'm involved with Jennifer O'Hannon," Jacob said.

Ozzie put his beer down. His eyes closed tight. He shook his head.

"You do understand don't you?" he asked.

"They will kill you. You *do* understand that don't you? Did you look at the photos?"

"I can count on you Ozzie."

"I'm going to tell you one more time Jacob. In this business loyalty does not cut both ways. The needs of the State are paramount. If things go wrong…"

"You told me."

"You're leading a fast life Jacob. Watch your back. We won't always be around to pull your fat out."

Jacob sipped his beer. "If things go bad here I have a place to jump."

Ozzie pursed his lips. "Dual citizenship?"

"Uh huh."

"That's smart, but remember Jacob you can run to us, but you can't run from us."

Ozzie put his beer down and got up. Jacob sat with his beer and watched Ozzie walk out. The following day the photographs that Ozzie had given him in the library of Bobo killing Matthew and hacking him up into mountable chunks were neatly laid out on Jennifer's bed.

"It's like this Jennifer," Jacob said. "Industrial espionage is big business. Governments spy and companies spy. Everybody spies on everybody."

"Espionage is a crime!"

"So is murder!"

Jacob sat on her bed kicking off his shoes and he made himself comfortable. Jennifer circled around him warily, her lips pulled back from her teeth, as if she were ready to bite.

"The full report will reach the authorities within an hour if I disappear or die," Jacob said. He grabbed up a handful of the photos.

"Your ex-beau was drugged and damn near dead by the time your bother got to him."

Jacob stared at her. "And that thing with the Chapel was cute. The chumps think they are going to get married and instead of you they get Frank and the crew. Nice!"

"Are you blackmailing me?"

Jacob lifted his head up off the pillow staring at her. "Damn right. You little tart! And don't forget it! If anything happens to me the truth about Matthew will hit the fan so fast it'll make your toes curl!"

"You can't…"

"Yes I can!"

Jacob leveled a steely gaze at Jennifer, who stood in the middle of the room, her face blank and her jaw slack. Her limited mind vacillated back and forth on what action to take.

"You look good like that bitch," Jacob said, "And oh, by the way get my slippers and get me a drink. The best damn bourbon Frank keeps around here."

When Jennifer stood there gawking Jacob yelled, "Move it!" Jennifer opened her jaw wide open as if to yell, and then snapped her mouth shut as Frank appeared in the doorway. His sad gray eyes moved slowly from Jacob to his daughter.

"Do it," he commanded, "The man wants a drink."

Jennifer rushed out. Frank leaned up in the doorway.

"You're probably what she needed all this time," Frank said.

"So long as she understands the rules of the game. I say and she does," Jacob said.

"Don't worry about her," Frank confided, "I know that girl and believe me she's a good sport."

A week or two went by. Jacob settled into the O'Hannon mansion. Jennifer, he came to realize wasn't even a good one-night stand and she certainly wasn't marriage material.

She was fallow. Not because she was a murderess, but she was without seed and, if all that wasn't bad enough, she was positively lousy in bed, but Jacob thought he understood her.

When Jennifer was planning to murder her lover she was a tigress. Jacob supposed that, in her thinking, it was a matter of equity, of value given for that she was about to receive. Tit for tat. But now with the big deal off she was limp as a rag.

Still, there was the lifestyle. Post-Grad education in biochemistry? Who needed it? Working was for chumps. Frank O'Hannon kept nothing but the best. Jacob settled in and getting Jennifer used to cleaning up after him. The best thing about his lifestyle, however, was the booze.

"Put the bill on Frank's tab," he said to the clerk. Jacob hefted a box of bourbon into his arms and started out of the liquor store when a voice startled him.

"Jacob?"

He turned around to see Ozzie standing in a shadow.

"Ozzie? I didn't see you."

"We gotta talk."

Ozzie came forward and took Jacob's arm and led him out of the store. "You remember what I told you?"

"Sure."

"Look Jacob," Ozzie said, "Move away. Change your name. Get a new identity. Maybe we can help with that."

"What?" Jacob asked. "Things are smooth," and then he checked himself thinking of another possibility. The thought crossed Jacob's mind that there really isn't any honor between thieves. Ozzie had helped get the goods on Frank, but now maybe the boy needed a little cash?

"You shaking me down?"

"No no," Ozzie stammered, "Nothing like that, except…" He stopped in mid-sentence and seemed confused.

"Ozzie?"

"Okay look Jacob," he said weakly. "Look, everybody is spying on everybody…"

"Yeah I know," Jacob said grinning. "It's what makes the world go round."

"Don't make cute with me!"

"Sure Ozzie."

"The FBI turned me. They set up a sting and caught me and a bunch of us with our hands in the cookie jar."

"Ouch," Jacob said still not catching the implications of what Ozzie was telling him.

"A decision was made."

Again Ozzie stopped speaking and let his eyes meet Jacobs. "A decision was made from on high," Ozzie said. And this time when Ozzie stopped talking Jacob's mouth fell open.

"You sacrificing me to save yourselves?"

"The police know about the murder of Matthew. They know what Frank O'Hannon does for a living."

"And?" Jacob asked.

"And they know that you had evidence of the crime. They know that you knew and covered it up. They know that you are blackmailing that tart Jennifer. They know the whole story."

Ozzie raised his hands up defensively as Jacob dropped the booze. Eight bottles of high price bourbon shattered on the sidewalk.

"Why?" Jacob shouted. "You bastards are throwing me to the wolves?"

"That's one way of putting it," Ozzie said defensively.

"Why?" Jacob shouted. "Damn it all to hell *why*?"

"Because you're expendable!" Ozzie shouted.

"We have ongoing work here. That crap for you was a favor! It was payback for the work you did for us. I'm sorry Jacob. Really."

Ozzie hurriedly walked away. Jacob watched him glance over his shoulder. Well Jacob thought, as he stood over the broken bottles of booze, at least he had been warned. Maybe he'd have time to get away. First though he'd have to get to the bank and…

"Mister Ebbtide?" a man standing in shadow asked.

"Jacob Ebbtide?"

He was wearing a business suit and held a badge in one hand and a gun in the other. Then Jacob saw others like him, carbon copies, all with weapons and badges.

"Don't!" he commanded as Jacob started to run. When Jacob stepped over the broken bottles, approaching the officer, he saw the handcuffs.

"You are under arrest."

That night was the worst in Jacob's life. The jail was over-crowded. It stunk, and what made it seem worse was that a lot of bums in the tank seemed positively pleased to be standing in puddles of their own urine.

The following morning a young thin muscular woman with mousy hair came into the tank. She seemed to know many of the bums and other inhabitants of the police station.

"Mr. Ebbtide?" she asked.

"Yeah," he said sourly. "Can you get me out of here?"

"I'm Gladys Hines," she said, "I'm your lawyer. I was retained by Frank O'Hannon."

"Frank hired you?"

"I just said so."

"To represent me?"

" Are you dense?"

Jacob laughed at this cute tough lawyer. And so, with that chance meeting, fate put another card on the table and three nights later Jacob laid his lawyer.

The trial for withholding evidence never took place. Gladys got the charges dismissed, although the evidence was overwhelming. Her obvious abilities as a lawyer excited him as much as her tight little ass.

It might have been Gladys. It might have been the Israelis. Gladys wouldn't talk about it. She laughed it off. After that Jacob went back to work in a research facility, but after a couple months, still itching for action and not harboring any bad feelings he had gone back to work for the Israelis. Blood, he always said and actually believed, is thicker than water.

Jacob liked passing information to the Israelis. To subsequent accusers he merely explained that it was all a matter of priorities. "Everybody is equal," he said half-jokingly, "but some are more equal than others."

"You place yourself in fast company," the Sheik said.

"It's an observation."

"Yes," Jacob admitted, "I was young. Going back to my old habits was stupid."

"Are you a smart man?"

"Smart?" Jacob asked. "I like to think so."

"I don't mean to offend," Kasan replied, "It seems to me, Doctor Ebbtide, that you get into situations as they say and then, well, you get out-smarted. People seem to, how do you say, see you coming?"

Jacob laughed. "Yeah, that's what they say. They see me coming." Jacob studied Kasan's face and watched his mouth relax into a smile.

"So," the Sheik said, "Going to America would not be very good for you. Or going back to Israel."

"But going to the Kingdom is worse," Jacob said.
"I like to play in fast games, although," he shrugged, "I guess I do get out-smarted. Still, in the Kingdom, the rules are very strict. In the UR I would have got six months jail time on that beef for withholding evidence. For the same crime in the Kingdom…"

"In the Kingdom," Kasan said, "You would be whipped. Then you would go to prison. Once in prison you would be whipped again. And, your prison term would be a lot longer than six months.

"We have ways to protect you. Even from yourself. We need you and all of your work will be shared equally with Israel."

"Is that agreed to?" Jacob asked.
"I speak for Sheik Gilla and for Sheik Abud.
"We will sign an agreement. Why is there a problem?"

Yes, of course, Jacob thought, the Kamsi would share with Israel. Why not? Once word got back to the Americans that he was working for the Kamsi there would be gold rush to strike deals with the Arabs. The dam would burst. Leviman and the Americans would be elbowing each other out of the way in a frantic effort to make deals.

Jacob put his face down in his upturned hands and rubbed his eyes. What the hell, yes, he could be useful in the Remnant, but then he'd have to deal with the legal system. And in the Remnant justice really *is* blind.

Scientists would want to talk to him about his work. They would want to know how he broke the satellite programming, but Gladys would want him behind bars.

"Aw hell," Jacob said out loud, but the Sheik put his finger to his lips,

"Doctor Ebbtide," Kasan said, "I feel we can work together. I believe it, but please do not swear."
"Sorry, but I don't think we can work together," Jacob said.
"I don't feel right about this. I can't betray Israel."
"But you wouldn't be," Kasan insisted.
"You'd be working for them at a distance."

"I'd be taking your money."

"But you would not betray Israel. I swear it," the Sheik said, but still Jacob understood that the Sheik was a trained psychologist, a skillful recruiter. He was a 'headhunter.' And Jacob also knew that going deep into the Kingdom held very real dangers.

Jacob stood up, met Kasan's gaze and the Sheik stood and walked to him. They shook hands and walked together to another room.

They left the interrogation room and entered a large dining hall decorated manner of a Kamsi palace: Fountains splashed water; the floor and walls were inlaid tiles of blue and aquamarine in elaborate designs.

A huge table of food was set out. Dates, figs, meats and loafs of many different breads. It was a feast. Jacob counted twenty chairs in the room and remembered, once again, why he didn't trust the Kamsi.

Now, however, with the interrogation over Jacob sat at the end of the table with the Sheik at the other end. Kasan made it clear that helping Israel was considered good business.

"Thank you," Jacob replied, "I don't know what I am going to do, but I am devoted to my country."

"America?"

"Israel."

"Yet you have decided to go to America?"

"It's my best choice now," Jacob said.

6

"Hear ye, hear ye," the Bailiff chanted, "Superior Court for the Seventh District of New Boston is now in session. The Honorable Gladys Hines presiding."

"Gladys?" Jacob asked disbelievingly. "Gladys?" he asked again.

When she appeared out of a side door and climbed the stairs to the bench Jacob stared with his mouth open in abject bewilderment. His ex-wife, the one-time criminal lawyer, was now a judge. Even worse, she was *the* Judge at his hearing. She settled herself, fluffed her black robes and smiled down at him.

"How nice to see you Sir."

Her smile faded. She busied herself with papers and then smiled down at him again.

"You have no idea, Sir, how nice it is to see you."

Her skin was rougher than he remembered. Oh well, Jacob thought, fifteen years will do that. Her hair was up in a tight bun, but the white still came through in thick strands, weaving in and out of mousy brown hair. When she looked down at him she seemed to be pointing with her straight nose.

"Damn," Jacob said regaining his composure.

"Sir?"

"Nothing, Gladys," he said respectfully.

"I forgot how straight your nose is."

A murmur of giggles echoed through the courtroom. Gladys pointed her nose at one group or another and the giggling faded into obedient silence.

It was all like a bad dream. Jacob had been arrested at Logan International Airport nearly the instant he had stepped foot on Remnant soil.

The Long Arm of the Law had scooped him up like so much horsepucky and deposited him in a New Boston jail. He had taken a night flight from Germany in the hope of getting out of the airport under the cover of darkness. He had almost made it. He stopped to glance at the newspaper headlines: 'New Boston gets New Mayor Today.'

He was putting cash into the newsstand vending machine when the tip of a very large handgun, with a dark shiny blue barrel, nestled snuggly into his right ear.

Obviously law-enforcement had been tipped off. Probably tipped off by Leviman, but maybe also by Sheik Kasan. Hell, Jacob thought, maybe by both. His attention went from Gladys to the courtroom.

The Courtroom was small, but well appointed. The Judge's bench was polished walnut. Ceiling spotlights shone down on it so that a bright yellow haze shone into the eyes of anyone looking up at it.

Jacob felt his eyes burning as he glared up at his ex-wife now Judge Hines. He wanted to rub his eyes, but the chain connecting his wrists with his shackles was too short. He would have to bend over, as if in a deep bow to her, to reach his eyes. He was not about to suffer that indignity so he blinked at her.

"You can't do this!" he protested. "Gladys," he insisted, "You should recluse yourself! This isn't right! You can't preside at my trial. You are prejudiced against me."

Gladys gaveled her courtroom to silence.
"The Clerk will read the charges."
"Felony flight to avoid prosecution."
"Gladys!" Jacob shouted, "You know I didn't…"
Again Judge Hines gaveled her Court to silence with a stern warning to the defendant. A Bailiff came up behind Jacob to insure he maintained proper decorum.

Jacob stopped squinting at the bench. His gaze lowered to the heavy iron shackles on his wrists and ankles. He looked over, to a long, raised, very attractive wooden ledge filled with potted plants.

Broad-leafed dark shiny green leaves grew in profusion. He thought of Solbean drying in the fields. Well, at least these plants were well cared for. He hadn't remembered Gladys caring at all for plants, but she had her courtroom filled with them.

"Felony drug possession in the second degree," read the Clerk.

"Not guilty!" Jacob shouted. "I'm not guilty!"

"Silence!" she ordered. "You were already found guilty of all charges fifteen years ago! You were tried in absentereo and found guilty. This is a sentencing hearing! You are being sentenced!"

"Gladys!" Jacob shouted, "I should be able to face my accusers! This is America! Isn't it?"

"Not anymore," she said officiously.

The Bailiff stared into space apparently, Jacob thought, to avoid blinding himself by staring up at the polished walnut bench.

"You sacrificed your rights years ago, Doctor Ebbtide. I am merely imposing sentence that was lawfully determined. Any court in the country could hand down this sentence."

She lifted papers. "It's all here. You can get counsel in jail, Sir. You can appeal your original conviction on all counts of the Indictment."

She hesitated and then added, "But of course, the fact that you ran is prima facie evidence of guilt."

She shuffled more papers and then raised her head.

"And I believe that the Statute of Limitations on appeals for a drug conviction is seven years. Your flight to avoid prosecution will be hard for you to explain."

Her green eyes opened wide and the corners of her mouth tucked up as her gaze seemed to take in the whole courtroom. "I believe, Sir, that train has left the station."

Jacob put his head back, straightened his shoulders and brought his tired eyes up to Gladys.

"Question," he said and when Gladys nodded, he asked, "What do you mean any court?"

"As I said, Doctor Ebbtide. Any court that has you can impose sentence. You could have flown into New York. They might take you or they might not. Prisons are filled to overflowing. Any jurisdiction, including mine that has you and wants to keep you, can impose sentence. That's the law."

"But it can't be right!" Jacob protested. His shackled hands were out-stretched as far as the chains allowed. His eyes were weepy from the bright yellow light blasting off the bench.

"Five years!" Judge Hines said. Her gavel fell with a thud against hard wood.

"This can't be right!" Jacob shouted again as he was dragged out of Court by the Bailiff pushing him hurriedly along to make time for the next case.

"Listen I'm telling you," he begged the Bailiff,

"I was married to her. I skipped out fifteen years ago. Gladys has it in for me. I skipped out with our daughter Talya and...and..."

"Shadup!"

The Bailiff shoved Jacob along a corridor. They passed connecting offices and hallways. When they came to a rear building door the Bailiff keyed his shoulder radio.

"Ready for pickup."

Moments went by as they waited by a steel door and then Jacob heard a vehicle brake to a stop outside. The Bailiff unlocked a heavy metal door. He took Jacob by the scuff of his collar and shoved him outside.

A small white prison van waited with motor running. A stiff breeze carrying rain came around the corners of the courthouse building. It fell in sheets down along gray granite block walls. Jacob hesitated to breathe the air. He turned his face up to taste the rain, letting the cool drops splash on his face. Jacob felt life and energy in the water. Spring was coming, but Boston still had a few weeks to go.

A shove from the Bailiff brought Jacob into the present. Jacob laboriously crawled into the vehicle. The Bailiff reached in and locked the shackles to the van floor. The vehicle door slid shut with a thud and locked.

Jacob tried to raise his feet to press against the partition facing him, but his shackles were locked to the metallic van floor. A cop turned around to him.

"Pipe down back there! If you give us trouble you'll wind up in Stonegate!"

Jacob stopped fussing long enough to stare at the cop. "What in hell is Stonegate?"

The van made a tight U-turn and headed down a long rain swept road between gray stone buildings. An iron gate opened as the van approached. A brick road led out of the Courthouse complex. The van made a right turn driving down the block. Jacob recognized the neighborhood.

It was an exclusive fashionable part of Boston with stately homes lining both sides of the road. The houses were old, but exquisitely maintained.

Dark green lawns, trimmed and manicured, swept down graceful landscaped visas to the road. He had not remembered this neighborhood as this pretty. Fifteen years before, if memory served, the houses were in dire need of repair and most had been partitioned for renters.

A quarter-mile later the van pulled up to a classic multi-story house. It had probably been built around the turn of the Twentieth Century. Obviously no expense had been spared in its construction or its renovation. Jacob vaguely remembered this section of town. He looked up to see the van stop.

An ornate wrought iron gate opened mechanically. As the eight foot drive gate majestically swung open Jacob caught sight of a dark green lawn, edged and maintained to perfection, spreading out under London Plane Trees; and then, masses of dense shrubby framing the dwelling house in a perfect blending of architecture and nature. Jacob marveled at the landscaping.

It would take an army of slaves lavishing time and sweat to maintain it. The recent rain added a mist, a gloss, to the architecture that reminded Jacob of the subdued quality of old paintings by the Masters.

Paintings so expensive that more than mere money was needed to acquire them. A man's minor daughter would have to be thrown into the bargain

The van pulled to a side door under a bright maroon canopy. Rain splashed off the awning onto the van's roof. The driver honked the horn once. The side door opened and two uniformed women stepped out onto the driveway.

One was large. She stepped forward and swung open the van door. She quickly undid Jacob's shackles and with a fast yank of her thumb motioned him out.

"Kom!" she said in rough German. "Ous!"

From his limited knowledge of German women Jacob's impression was that this woman to be a winner of an East German beauty contest: Who among the contestants looked most like a violent car wreck?

Yes, he thought, she would win. Jacob wondered idly if she had taken some sort of steroid therapy to gain such hairy muscle mass.

The second woman, also in a white uniform, was petit and even smaller than Talya. Her hair was reddish with a blond streak curving down over her right ear, folding over her shoulder. She remained respectfully by the door, holding it open while Car Wreck approached the van. Jacob noticed, from the van window, that the big woman had a handgun strapped to her fat waist.

A cop got out of the passenger seat and handed a clipboard to her. She reviewed it, peeked inside at Jacob and nodded. "Sign for him!" the driver yelled over the din of rain hitting the awning. Car Wreck signed.

The slightly built guard with the blond curl came forward without command and grabbed hold of Jacob's shackles and started for the side door.

Her name badge read only, 'Maria.' She pulled on the chain connecting Jacob's wrist shackles with his ankle shackles and didn't give him a chance to see Car Wreck's nameplate.

Jacob stood by the side door with the petit attractive cop in the doorway holding the chain. When he tried to turn around to see behind him she tugged on the chain.

Jacob faced around and followed her inside without protest. He heard the uniformed cops get back in the van. Behind him the van pulled away and he got a glimpse a small landing with a bathroom to his left. Straight ahead from the door stairs went up to what Jacob thought must be a kitchen. The smell of hot bread baking whiffed down the stairs toward the open side door.

"Keep moving," Car Wreck said. Maria went quickly down a flight of steps to a basement. Jacob stumbled with the shackles, but kept his balance as he hobbled down the stairs.

At the bottom of the stairs the room opened into a wide basement converted into a makeshift medical center. A low white acoustic ceiling gave way at the walls to bare white tiles, brown file cabinets, carts on wheels stacked with medical paraphernalia, needles, canisters, packages of bandages, free-standing medical test equipment and then, over by a back wall, Jacob saw an examination area.

"Over there," Car Wreck gestured to a padded bench. "Sit," she commanded and Jacob sat on the end of the bench and gazed around.

"Medical exam?" he asked.

Car Wreck stood back with her hand on her handgun holster while Maria undid his leg shackles.

"Get them boots off," Maria said pointing at his prison issue shoes. Jacob slowly rose one foot, sliding one shoe off, then the other.

"Now the socks."

Without hesitation Jacob did as he was told. The socks tumbled down to the floor on top of the boots. Maria stood back a couple feet while both women stood looking intently at his feet.

Jacob stared back at them, looking from them to his feet, and then wiggled his toes. He lifted his shackled hands in a gesture of what gives?

"Size ten and a half?" asked Maria.

"More like eleven," said Car Wreck in English.

"Eleven wide," Jacob said. "I've got big feet."

Car Wreck leaned forward looking carefully. "Ya," she said straightening up. Her forehead furrowed in thought.

"Gross," she said.

"I think so too," Maria answered.

Jacob watched her go to a medicine cabinet. "Excuse me," Jacob said politely, "but my feet are bigger than that. I'm a scientist. I know the metric system very well."

Both women ignored him as they rummaged through the medicine drawers finally bring out a large bottle of pills. Maria came over and dropped two huge pink pills into Jacob's upturned hands. With it she brought a small paper cup of water.

"Is this a medical examination?" Jacob asked while studying the pills.

"Only for your feet," Maria said.

"Take now!" Car Wreck commanded.

Jacob put the pills in his mouth and then took the offered cup of water. Within moments, as both guards watched him, a warm glow quickly settled into Jacob's stomach and spread to his head. Maria came forward and propped his feet up on a table extension.

Jacob leaned all the way back and laid out flat and in a moment he was peaceably asleep. He did not know how long he slept, but he felt cozy. Warm, billowy folds of soft cloth enfolded him.

Sounds of typing on computer keyboard filtered over other noises: an overhead fan, someone walking on wood floors, a murmur of voices. To his right a man cursed under his breath. Jacob opened his eyes. Soft, pleasant, subdued lighting illuminated a deep rosy gloss of stained wood paneling.

His gaze took in a loft that was, to his estimation, fifty feet long by about forty feet wide. Ten feet from him wood panels extended from a solid wood floor up to clerestory windows high above where the walls gave way to a slanting roof, also of bare wood. Skylights far overhead glistened. He could make out raindrops sliding down dark skylight glass high above.

His arm flopped out of a sleeping bag. Jacob rolled onto his side, propping his head on his upturned hand. With a sudden shock he realized his wrists were not shackled.

He sat up, throwing the sleeping bag off his legs and looked at his feet. He wore white socks. His ankles were free. There were no chains on him. He looked up to a post by a door to see the shackles hung on pegs. Jacob sat up, crossed his legs, and looked around.

He felt good except that his feet felt like little stones were rolling around inside of them. Rubbing his feet he felt tiny round objects under the balls of his feet. They seemed to roll around under the skin like ball bearings. He was about to ask about his feet, but glanced up to see both of the female guards busy at work behind a computer console.

The guards worked at a workstation at about the two o'clock position. Jacob watched them for an instant and then lowered his gaze to a plush red carpet, a carpet runner, extending the length, left to right, of the whole loft and out to a single door on his left.

Jacob looked up when a man wearing only a bath towel came strolling out of another door in the wall to the right of the computer workstation.

Maria said something and the man stopped. Car Wreck punched some keys and made a gesture and the man walked into an area to Jacob's right.

The man came to a stop about ten feet from Jacob who blinked, watching wide-eyed, as the man let the towel drop to the floor. The man settled down onto a sleeping bag. Then, for the first time, Jacob noticed other men beyond him moving silently around as shadows in the dimly lighted loft.

Over a patter of conversation between Car Wreck and Maria, the whirl of ceiling fans, and muffled noises of the other men settling themselves in their sleeping bags, Jacob heard traffic on the street. Trucks rumbled by and car horns honked.

"Where am I?" Jacob asked.

Car Wreck made no comment, but picked up a telephone, dialed and said something that Jacob could not hear. She looked over at Jacob.

"Here we go," she said in English. "Get ice."

Maria looked up from the console, stood up and went back in the direction that the man with the towel had come and reappeared a moment later carrying a wooden bucket with a bright copper ring around the top.

As she came closer Jacob saw ice cubes piled up in a mound glistening in dim white light. Maria came around the computer console placing the bucket on the floor in front of Jacob. He leaned over looking and then put his hand inside scooping up a handful of ice cubes.

"Ice?" Jacob asked.

"Trust me," Maria said. "It's the best friend you've got."

"No talk!" Car Wreck said from the console.

Maria backed away and went back to her work alongside the Car Wreck. Jacob felt hot breath on his neck. A man next to him, that had dropped the towel exposed his privates and then leaned to him.

"Don't do it," he cautioned.

"Where am I?" Jacob asked again.

"Don't be stupid," the man said. "Really," he pointed to the wooden ice bucket. "See where it is?"

"What?"

"The ice," he said. "Make a mental note of it. You're going to need it."

"What? Ice?"

"Yes. Ice."

Jacob sat looking from him to the ice to the women at the computer console who showed no further interest in him. The man to his right was still breathing on him with his face too close and his eyes too intense. He was blond, youngish, muscular, well-proportioned and naked as the day he was born.

"Oh man," he whispered, leaning close to Jacob's ear, "Don't. Really. Don't play her game."

Jacob tried to ignore him, but wasn't used to naked men being so damn close. It was unnatural. The man seemed harmless enough, but was incoherent and maybe crazy. Ice? Why was a naked man babbling to him about ice?

Jacob looked away from the man and down to see that his own red jail jumpsuit had been replaced by a soft, loose fitting, one-piece garment that ended at his ankles. In the soft light his garment seemed earth-color green and brown. It was very comfortable. He breathed a sigh of relief that he was not naked.

"Believe me," the man implored, "Don't…"

"Kelly!"

The man slid away from Jacob retreating quickly to his own sleeping bag. He lay out on the floor like a big white dog, butt down and private's stiff. His legs flopped open at the knees.

"You need more work Kelly?" Maria asked him.

At these words Kelly slipped into his sleeping bag. He rolled up folds of cloth to his chin and closed his eyes.

Jacob watched all this with his mouth slightly open in amazement. Behind the man Jacob saw a small toilet, a washbowl, and a small bookcase with some magazines lying loose on top.

Beyond the man Maria called Kelly, further into the loft, others were settling down for the night. Jacob counted four others in the loft. Jacob noticed that his area on the floor had the same layout as the others.

Directly behind him was a toilet, a small washbowl, and a tiny bookshelf. There were no chairs, stools, or other furniture in the area. The ice bucket rested where Maria had placed it inside his area.

To his left the door opened and Gladys walked in. Both Car Wreck and Maria hurried around the computer console to stand respectfully at attention.

"Problems?" Gladys asked them.

They shook their heads and then faced front at attention with hands down at their sides. Gladys stood inside the door carefully surveying the entire loft area. Her gaze went from the computer console, to the carpet strip, to each of the men in the facility. Jacob sat passively watching and only when Kelly cleared his throat did Jacob look over.

Kelly was in his green single piece suit and stood opposite Car Wreck, his beefy frame upright, staring straight ahead and further down the line Jacob saw the others standing like soldiers on parade. Gladys stood by the door. Her attention now fastened on him. She cleared her throat again and Jacob slowly got up.

As Jacob got to his feet he shifted his weight, lifting his right foot to step out of his area, but stopped abruptly as Kelly harshly whispered, "No!"

"Put that man on report," Gladys said. Car Wreck nodded. "On report."

"I warn you!" Gladys said. She faced off in front of Kelly. "That infraction will cost you two more months. Was it worth it?"

Jacob watched Kelly slightly move his head in his direction as he answered, "Yes Judge."

"Do you want three more months?"

"No! Please Judge."

Jacob heard Kelly's sigh of relief when Gladys walked away from him. Gladys obviously terrified him. Jacob stood, looking back and forth from Gladys to Kelly and from Kelly to Car Wreck.

"Gladys," he said, again lifting his right foot to bring himself out of his area, out onto the carpet to confront her, but before he stepped Gladys raised her hand, stopping him and then gestured to the ice bucket. "Has it been explained to you?"

"Nothing's explained," Jacob said again leaning forward, but stopped again seeing Gladys' sudden interest in his feet. Her attention went from him to the ice. She walked stiffly up to him. Positioning herself face to face, she stared at him.

"It's a private prison," the Judge said her voice matter of fact.

"Crime is rampant. Spreading inmates out into dozens of private prisons makes perfect sense. Spread out the problem. This technology," she pointed at the red carpet, "is cheap and makes control of prisoners easy."

"Private prisons?" Jacob asked. "What's a private prison?"

"Homes like mine get restored," Gladys said. "We open our homes to the prisoners and they work off their debt to society."

"You live here?"

"Of course."

Jacob staggered back a couple feet with the implications of where he had suddenly found himself. "My god," he groaned, "*You own a prison?* You live here?"

"This is my home."

The Judge turned to Car Wreck. "Be sure he gets the basic instruction sheet." The Judge walked stopped again in front of Jacob's cell.

"Unemployment is at twenty-five percent; and yet," she gave Jacob a steely gaze, "It's impossible to find good help." Gladys looked over her shoulder at Car Wreck.

"Right?"

"Yes Judge."

"You are in prison Sir. You will do well to remember that." Her chin lifted quickly as Jacob was about to make a reply, but he hesitated unsure of himself. He stood looking from his feet to the ice.

"You remember the basement?"

"Yes. Is the basement here in this building?"

"Five floors down Sir. Your feet were examined." She pointed to the ice bucket.

"Special metal pellets were injected under the skin of the balls of your feet in the fleshy part of each foot. Each pellet is encased in a special heatproof plastic shell. The pellets are small, but you do feel them. Don't you?"

Jacob looked over at Kelly, who was standing at attention, staring across the room at Car Wreck who, in turn, was staring straight back at him.

"Well yes," Jacob said. "I don't understand. Why put metal pellets in my feet?"

"Don't be stupid Sir. I said it once. I won't say it again."

Jacob swallowed hard and stood staring back at her. Jacob wanted to tell himself that he didn't know this hard woman, but her expressions, her mannerisms, her strident personality hit him. Memories flooded back. If ever two people were mismatched he and Gladys were it.

"Tell me," he asked softly.

"If you step out onto the carpet," Gladys pointed to the six foot wide red carpet runner, "an electrical grid under the carpet will cause induction heating of the metal pellets in your feet. When you step outside your cell the grid heats the pellets. Your feet burn. The muscles in your feet burn from the inside out."

Jacob's eyes opened wide in shock. He blinked back sudden tears. Gladys stared at him indifferently. Her green eyes reflected subdued light giving her eyes and her face a gray pall.

As a scientist he was quick to realize how simple, how effective, such an imprisonment system could be and how horrible.

"You have no right to do such a thing!" he shouted. "Burning a prisoner's feet?" he leaned forward, careful to avoid the carpet. "You damn barbarian!"

Gladys didn't answer. Around Jacob nobody spoke, not the women nor the other prisoners. Gladys came up closer, almost nose to nose with him.

"You are in prison!"

When Jacob stepped back from the carpet by a few inches he saw her mouth turn up into a mirthless grin. "Good," she said. "We have communicated."

When she made a turn to face the women Jacob reached out, touching her arm. Gladys pulled away.

"That's an infraction!"

Jacob pulled his hand back. "I only want a word with you!"

The Judge looked up and down the line of men. Lifting her chin, she spoke commandingly.

"At ease!"

Kelly and the others slumped down to their sleeping bags. Car Wreck took a position closer to Gladys as if to protect her.

"Each cell can be electrically isolated," Gladys said.

"We can let you move around or lock you down Also different metals in your feet respond to different frequencies and levels of electricity in the grid. When we lock you down you must stay in your own cell. If you stray out of your area, even into another prisoner's area, your feet will burn.

"Believe it! We can let you move around or lock you down. How much freedom of movement you get depends on your level of cooperation."

Jacob decided to pay his last card. "I am a scientist," he said, leaning forward, but being careful not to step over onto the carpet. "My work is very important. I've done something to hurt…"

"Your work here is important," Gladys interrupted. "All of you have important work to do." She faced Jacob again.

"You, Sir, will pick snails out of my garden! For the next five years, ten hours a day and seven days a week. *That* is your important work!"

Jacob's jaw dropped. He found his glasses in a breast pocket of his prison jumpsuit, put them on, and looked Gladys up and down.

"You can't be serious!"

"I am very serious."

Gladys stepped aside as Car Wreck came up with a small pamphlet. She handed it to Jacob. He adjusted his glasses, but still had trouble reading the small print in the dim light. He scanned down through the document.

"What?" he asked. "Do useful community work? You mean like a chain gang?"

"You'll pick snails out of my garden," Gladys said again.

From someplace down the line of cells one of the prisoners grunted. Jacob noticed that they were all facing him, standing there, as if waiting for him to say something.

"The man has seen the light," one of them said.

Car Wreck stared over in them man's direction. He moved cautiously backward toward the rear of his cell. Car Wreck walked down the line of cells to the man.

"Would you rather be in da Big House?" Car Wreck asked. From around the loft the other prisoners muttered. Jacob heard a few chuckles mixed with hurried conversation.

"Quiet!" the Judge ordered.

Jacob read down the document. His eyes came to a stop at a paragraph entitled, 'Restitution.'

"And I will have to pay you forty thousand UR dollars per year to cover the expensive of housing me?" Jacob asked looking up from the material.

"This is a business," the Judge said. "Not a charity."

"This can't be legal!" Jacob shouted. "Charging money to imprison me? What the hell!" he roared, "*It can't be!*"

"Enough! It's time you understood. You will spend five years working off your debt to society. Consider yourself lucky. You could be in Stonegate."

"Isn't there some authority I can appeal to?" Jacob asked. "Don't I even get a telephone?"

"Telephone?" Gladys asked with contempt. "You get two calls a month. Five minutes per call. Local calls only. Telephone calls are expensive. We add the charges to your bill." Gladys gestured to the printed matter in his hand.

"And if you screw up here Jacob I will personally sign the order that puts you in that radioactive hell."

"What?" Jacob asked.

"Stonegate. People go in, but they don't come out alive."

She poked him in the chest. "You'll pay for your crimes, Jacob. And, you'll pay for taking my daughter." Gladys leaned closer.

"You are going to pay. Every day. Every hour. Every minute!"

"It was best for her," Jacob insisted. "I swear it. I took Talya for her own good. I think you know it!"

"The Court gave me custody."

"You rigged the syst…" Jacob let his words hang in air as he looked into her hard eyes and wondered how he could have ever loved this woman.

"Gladys," Jacob said plaintively, "You know me better than that. I've never physically harmed anybody in my life."

"No?" she turned to him again. "And kidnapping?"

"Gladys…" Jacob started to protest, but she held up a hand.

"I can't prove it Jacob, but believe me if I could you would be in Stonegate."

"I took Talya," Jacob admitted, "but I wanted her to have a good childhood."

"She would have had more opportunity here!"

"You were a mob lawyer!" Jacob shouted. "She would have grown up on the streets!"

"That's crap and you know it!" Gladys shouted. "Besides the Court gave custody to me!"

"She's missing. Did you know that?" Jacob suddenly felt a terrible fear, a suspicion, welling up in his heart.

"Do you know where she is?"

Gladys was mob connected. The entire world was corrupt. People on the wrong side of the law often knew things that honest people never figure out.

"Do you have information about her?" Jacob asked.

"I know she's missing!" Gladys said. "I talk to Leviman. Our daughter is missing thanks to you!"

"I had nothing to do with that!" Jacob shouted. "I'm worried sick about her!"

"You'll pay," Gladys said taking a step away from him to effectively end any personal communication. She moved with the same tough, muscular movements he remembered.

Time was when his body responded to her thin muscular legs, but now Jacob wondered if maybe he could make it away from her and her prison by crawling on his hands and knees and then reality hit him. He felt weak and empty inside. Jacob hated the idea of crying, of breaking down, in front of Gladys, but the whole damn thing was too much. He was not only a prisoner, but also *her* prisoner in *her* prison.

"Gladys," Jacob said desperate to find some agreement with her, "I wanted…"

"Gladys," his hand reached out, but didn't touch her, "I came back to help my country. I've done something wrong and…"

"Lawyers for the Government will be here someday," she said. "Don't worry about that. After you serve your time here. All five years of it. The Feds might still want a piece of you. I will turn you over to them." She gave Jacob the same cold smile.

The Judge motioned to the women. "They are Trustees so don't get any notions about harming them."

She briskly walked over to Car Wreck taking the clipboard. Gladys reviewed a document, nodded and walked out without looking at Jacob. A steel security door wheezed closed behind her.

Within minutes of Gladys leaving his anxiety had given way to exhaustion. The evening and night went by uneventfully. The drug made Jacob sleepy. He slept until he heard Car Wreck's loud voice. He opened his eyes to see light streaming into the skylights far above him. Car Wreck came forward and raised the clipboard. She gave the work detail for the day.

"Kelly, Ebbtide, and Justin. Your detail will be the right side lawn. Hand mowers only. The right side lawn must be mowed and trimmed!"

Justin moaned.

"Oh god!" he said. "The whole damn thing?"

Car Wreck set the clipboard down on the computer console. "You want I should write you up?"

"No," the inmate said sulking.

"What?"

"No. Please."

"Shay and D'Lill. You will do City work!"

Moans came from the two men. "Can't we pick up snails or mow grass?' one of them asked.

Car Wreck walked around the desk. "Work details have been approved."

"I hate picking up city garbage!" Shay said.

"Sanitation workers get wages! You on the other hand, Shay," Car Wreck pointed a beefy finger at him,

"Have committed crimes. You will pay for your keep by doing work.

"Take a tip from us. Be a trustee. You get paid. You get privileges. We go home at night."

"I got four more years of this slavery before I qualify!" D'Lill shouted.

"Quiet!"

Car Wreck went behind the workstation and made a call. About five minutes later two unformed policemen came into the loft. Car Wreck gave them the clipboard and they motioned for Shay and D'Lill to follow them.

Jacob watched the men leave for the day's work on the garbage detail. He wondered what his day would be like mowing with Justin and Kelly.

Ten hours later he had the answer. At the end of an exhausting day Jacob came back into the loft with the others. He was too tired to think of anything and slumped down, collapsing, into his space with his eyes wide open. Car Wreck bent down to get something under the counter and stood up with a stack of five rolled white towels. She put them up on the console. "Showers! Two minutes each. Justin is first. Move down da line!"

Jacob sat up. Crossing his legs, he watched Justin stand at the edge of the carpet. Maria gestured at him.

"Grid is off! Stay in da lane!"

Justin came out naked and walked confidently up to Car Wreck who gave him a quick look up and down and handed him a towel.

"Two minutes."

Justin took the towel and walked out of sight. With Justin gone to the showers there were only four prisoners left in the loft. Jacob looked over to see Kelly nudging up closer to him again. This time Jacob wanted to talk. In ten grueling hours outdoors he never got within talking distance of the others.

"Kelly?" Jacob asked extending his hand. "I'm Ebbtide. Jacob Ebbtide."

"Yeah. We know about you. We didn't get a chance to talk outside."

"That's slave labor," Jacob said. "I never knew people could work so hard."

His arms and hands trembled as he tried to scramble out of his prison jumpsuit. The clothes hit the floor. Jacob kicked the jumpsuit into a corner.

"It's rough now, but it gets worse as the weather heats up. Summer is hell. Real hell. You'll see." Kelly said.

"This will kill me. I'm a scientist. I can't take it."

"You can take more than you think Ebbtide," Kelly said. "Besides what the alternative?"

Jacob hung his head for a long moment. He let air fill his lungs. They sat for a while. Neither man spoke. When he felt better Jacob looked to Kelly.

"You knew my name?"

"The Judge said you were coming. You're a feather in her cap. You must have some political friends. I mean, you know, to get in here."

"How's that?"

"Most go to Stonegate or other such hellholes. These micro-prisons are tough to get into. Boutique prisons like this have a waiting list. Prisoners pay bribes to get in.

7

"Consider yourself lucky." Kelly said. "But I guess, er, there's something between you and the Judge. She's fair, but she can be a bitch."

"We were married for a year. We had a daughter," Jacob said.

"Married?" Kelly exclaimed blowing air. "Man. That's tough. Damn."

"When I knew her she was a mob mouthpiece."

"You ran with that crowd? Kelly asked. He sized Jacob up and down. "You don't look the part."

"Appearances can be deceiving."

"Don't I know it."

Kelly flopped onto on his sleeping bag. He wrinkled his nose. "They're all on the take one way or another. The government, the lawyers, the mob. No difference."

"You should know," Car Wreck said.

Kelly sat up, getting to his knees and then to his feet. He dragged his sleeping bag across the floor to the side of his area. He sat down and crossed his legs across from Jacob.

"You queer?"

Jacob winced.

"Good," Kelly said, "Me neither."

He indicated the others further down with a motion of his head. "Look," Kelly leaned closer, whispering,

"I'll protect you, but you know in the world today nothing is free. Right?"

"I can take care of myself."

"Not if the three of them come at you."

Kelly indicated the two trustees. "You see the grid is out there." He pointed to the carpet. "Not in here. Yeah, they can lock us down, but mostly they don't. Remember trustees go home at nine p.m. They turn the system on and leave. We're not going anywhere. The system is escape-proof."

"No prison is escape proof."

"This one is," Kelly said, "unless you find a way to fly," he looked back at the others, "And on cold winter nights," he gestured, "They are going to want company. They are going to crawl into your sleeping bag."

He came closer whispering, "After a day of hauling garbage, they'll feel ripe and want service! Get it?"

Jacob pushed Kelly away. He pointed at the small wooden bookcase.

"Kelly I'll tell you," and raised his voice so the trustees at the computer console and the other prisoners could hear, "Anybody coming into my space at night gets this bookcase wrapped around his head."

"Okay," Kelley said, "but everybody needs friends. You remember that." He spread out on the sleeping bag. "This is a barracks. People get to know one another. It pays to be friendly."

"I'm not that friendly."

"Yeah fine," Kelly persisted, "but we don't get any women in here. As the months drag by you might change your mind."

Jacob looked up to see D'Lill coming back after a quick shower.

Jacob leaned over to Kelly while watching the procession of men going to and from the showers. "Why not take a hike when we are outside? I mean collect the garbage? Get behind the wheel of the garbage truck and go!"

"Why not?" Kelly asked.

"Yeah," Jacob whispered. "Breaking out can't be that hard." He paused adding, "I mean, you know, when you are already outside. I can't see where there's a problem."

Kelly lay flat on his back, lifting the soles of his feet up. "Pellets," he said. Jacob saw Car Wreck looking over at him and Kelly cringed. She seemed to know they were talking about breaking free.

"It's all done from here Ebbtide," Car Wreck said. "Where you can be. If you wander anyplace out of your area your feet cook in your shoes!"

"But I'm a scientist!" Jacob objected pointing to the carpet. "You can't have a power grid covering every square inch of earth!"

"No need," Car Wreck said. "Here and there. Bus stops, rest rooms, sidewalks. All hidden. You won't know until you're on the ground screaming."

Jacob gulped. So that's how they did it. It was a random placement of the grid. He looked up at Maria typing at the computer.

Sudden understanding dawned on him. They could work the whole system right from the computer. It worked like a telephone network, but in reverse. If they dialed up a sector, the electric grid was switched off. But, if an inmate went outside an authorized area, if he made a run for it, he would have to run on his knees. *Or steal a car?*

Car Wreck stood behind the computer console still looking at him. "Five years, Ebbtide, but ten if you steal a vehicle."

"I knew a guy what stole a car," Shay said. He came to the edge of his cell joining in.

"With gas rationing you can't get far. And step out of the car at a gas station?" He winced.

"Five years for the break," Car Wreck said. "Five more for stealing a car."

"Is the whole damn country wired?" Jacob asked looking from her to Kelly to Justin.

"Naw," Shay said. "It's like landmines. You don't need them everywhere only where soldiers are likely to walk. Sure. You might be able to walk away, but you're taking a chance."

Jacob was quiet and finally realized that breaking out wouldn't be so easy. It was those damn pellets. He rubbed the balls of his feet and could feel them rubbing under his skin. Hmm, maybe he could cut them out with a knife? Hadn't the Car Wreck said something about a hypodermic needle in the basement?

"Is it sinking in Ebbtide?" Car Wreck asked.

"When can I make a call?" Jacob asked changing the subject again.

"Two calls a month," Car Wreck said. "Local only. You're allowed one call now."

When Jacob looked up she tossed a cell phone to him. His arms trembled as he reached for the phone. Just as he caught the phone she tossed a ragged New Boston phone book.

While the other prisoners went to the showers Jacob flipped though the phone book. It was old and torn with most of the listings removed. Only pizza places and others that delivered food were listed along with a few other professional listings.

"We can order food?" Jacob asked.

"If you pay for it," Car Wreck answered. "Tell me or Maria if you're going to pass on a meal. Otherwise the charge goes onto your account."

Jacob moved around to get better light on the smudged pages. He leafed through the almost useless phone book for a few minutes.

There were no numbers for any government agencies and Jacob felt, without being told, that any phone numbers he might get for the government wouldn't work on this phone or on any phone he was likely to use.

He had one other hope. That was to make a local phone call to somebody that might remember him. Maybe get a message out to the Remnant government that he was in the country and willing to work. Maybe, Jacob thought, somebody would come to his rescue.

In a few moments of flipping pages Jacob found the listing for M.C. Handy the consultant. Jacob remembered Handy. He remembered a grizzled old fellow years ago that got stung by a laser-branding device that Jacob, as a student, had helped design. Jacob closed his eyes and put his head over as a deep sense of guilt welled up in his heart. He deeply regretted any effort he'd made to build that terrible thing. Now he wondered if Mister Handy was still alive and doing well. He dialed the number.

"Handy," a man's voice answered.

"Mister M.C. Handy?" Jacob asked.

"The same."

"My name is Ebbtide," Jacob said. "I think I remember…"

"Sure," Handy interrupted. "That smart kid. Student if I recollect. Thought some husband would have bagged your ass by now."

Jacob swallowed hard. "I'm sorry you remember me in that light," he said.

"What other light is there?"

"Well I called to see if you're all right. Not to pick a fight with you."

"I'm right as rain," Handy said. "Still in the same business. No thanks to you or Puller."

"Still holding a grudge about the laser?"

"Those spots don't wash off!" Handy said. "And if I ever get my hands on that jackass…"

"That was fifteen years ago," Jacob said. He looked up to see Car Wreck coming in his direction. Apparently his time was up.

"Listen, Mister Handy," Jacob added quickly, "I just want you to know that I'm in the Hines Prison House. People in the government might be looking for me and…"

"Hines Prison House?" Handy asked. "What…"

Jacob looked up to see Car Wreck with a small electronic gadget in her hairy paw. She cut off his call. Probably on orders from Gladys.

"It's easy to pick up time around here," Car Wreck said. "Asking for a lawyer and causing the Judge inconvenience can get you six months. Understand?"

"Yeah," Jacob said. "Don't try to reach out to get help. Gladys doesn't like it."

"And don't crack wise either," Car Wreck said. "You'll pick up more time."

Jacob handed the phone to her. Slipping it into a pocket, Car Wreck dropped a towel down to him.

"You're up," Car Wreck said as Kelly came back from the showers.

"Is everything done here naked?" Jacob asked.

Car Wreck looked away in false modesty, but touched the towel with a toe. Message received.

After his shower Jacob laid down on his sleeping bag. He was too tired to sleep and felt like talking.

"Are all the floors filled with prisoners?" he asked Kelly

"This building," Jacob asked. "How many prisoners in the whole house?"

Kelly sat up grinning. "He wants to know how many prisoners in the whole house!"

Car Wreck looked up. "Five including you."

Jacob felt time resting heavily on his soul. Stuck in a small patch of floor with four other lost souls for companions would kill him. Over in the next cell Kelly was getting comfortable.

"What's the score around here?" Jacob asked.

Kelly explained about the lights and the video surveillance system that were activated to alert Central Enforcement Command of any missing prisoners in any of the dozens of micro-prisons spread around New Boston. Any prisoner stepping on the carpet would set off the alarms and alert the authorities and the medics.

He knew all this he explained, because he had until recently been a Captain in the New Boston Police Department. His fall from grace, Jacob learned, had been fast and hard.

Kelly said something about others bribing their way into a prison house. It was easy time compared to Stonegate. Jacob felt that most of the men on the floor were low-level crooks, but something about Kelly indicated something different.

"By the way," Jacob asked, "What are you in for?"

"I killed my wife," he said. Kelly rolled over and faced Jacob. He murdered his wife and he had also at the exact same time murdered the previous Mayor of New Boston.

Jacob remembered reading the headlines of the newspaper about New Boston getting a new major. Reading that headline had cost him the thirty seconds he had needed to make good his undetected entry into the country. Jacob was thinking of 'what if' when Kelly slid closer.

"There was a hostage situation on the top of the Dime Building," Kelly said.

"The Dime Building? You mean the skyscraper?" Jacob asked.

"That's the one. I was a Captain in the New Boston Police." In the dim light Jacob saw that Kelly had slipped out of his sleeping bag again. His one-piece suit was crumbled on the floor by his toilet. Kelly stared at the ceiling and talked.

The Dime Building? There was nothing up on the roof but an observation platform. Although Jacob had never been on the observation deck he knew that people on top the Dime Building could look for miles around.

In Old Boston there was damn little to see, except in the south the Government had confiscated some of the old Irish neighborhoods and turned the whole area into Stonegate that doubled as a prison and as a radioactive storage dump.

Tens of thousands of vehicles, mountains of scrap building materials, demolished houses hauled in by boxcars. Trainloads of dirt had been brought in and spread over the mounds and quickly hydro-seeded to get grass growing over the waste in a futile effort to stabilize it and bury hundreds of thousands of tons of radioactive poison.

It rained. The soil didn't hold. Muddy, highly radioactive ooze flowed through the streets, downhill into other adjoining neighborhoods. Then, those other adjoining neighborhoods had to be abandoned because the toxic soil, and radioactive rainwater had washed into basements.

Puddles of radioactive ooze ruined the houses even though it was only the basements that were actually radioactive—and the roads leading to the houses.

The best view of the carnage caused by the consumption of radioactive gasoline in the Northeast was from the top of the Dime Building. The city center was hollowed out. Some hills, some grass, and many radioactive ponds mingled together forming Lil Po Lake.

People also visited the observation platform to see Stonegate that was located right in the middle of the radioactive dump. The prison was a makeshift collection of tin buildings built on concrete stilts.

The only way into the prison was by air and the only way out was also by air or a lead coffin. There were no bars on the windows. No locks on the doors. If a prisoner walked out or away from a work detail he was told to keep going. Prison personnel wouldn't go looking for him and wouldn't take him back. Every prisoner knew that he was giving himself a death sentence if he walked away.

"How can anyone have hostages up there?" Kelly asked.

"I donno," Lieutenant Allan said.

By the time they arrived with the bomb squad a crowd had gathered out in the street at the entrance to the building.

"We began crowd control," Kelly said. "I looked up, but couldn't see the top of the building. It was night and raining. The rain came down hard. Rivers of black water gushed off the building. People in the street were sloshing around in it. An old bum rushed up to our van."

"God," the old man cussed, "He's got winos up there. They're wired to explode."

The man stepped away quickly and looked at our SWAT van and equipment.

"How'd you get down?"

"Elevator," the bum said.

Kelly brought his face down to the man. "No," Kelly said, "I mean how do you know what's going on up there? Nobody else does. Why you?"

"The guy sent me down." He gestured at the observation platform far above. "He set me free."

"What's this guy look like?"

"He's old. Like me," the bum said. "He's up there now. Says he going to kill'em if he don't get want he wants."

"He's up there now?"

"He's got one of them devices." The man clicked his thumb and finger together.

Kelly looked up. Low clouds and rain obscured the building. The wind blew driving sheets of rain ahead of it.

"How many hostages?" Kelly asked.

"Eh?"

"How many people?" Kelly demanded.

"Six now," the man said. "Seven with me."

"Well you're not up there now so you don't count."

When Fire Battalion Chief Mullins arrived Kelly pushed the old man aside.

"We got a bomber," Kelley said. "Clear the building and assign an elevator for your people and one for us."

When the old man started to walk away Kelly grabbed him. "You're going to show us," he said.

Kelly and his crew entered the building. The bum was right. All the elevators worked. Kelly, Allan and two cops and the bum rode up together.

The elevator stopped two floors under the observation deck. The firemen locked off the elevator and stepped out. The hallway was dark. Even the red emergency lighting was out. Flashlights poked here and there in the gloom. From someplace above water seeped down the stairwells.

"He's got lights rigged," Mullins said.

When they got to a stairwell they stopped, because behind them in the darkness they heard the elevator moving.

"He's also got the elevators rigged," Kelly said.

"I had to climb down," the old man pointed up the dark stairs. "Two floors. Those elevators don't go all the way up."

"Wait here," Kelly said to the bum. "We'll take a look-see."

Kelly and Allan climbed up the stairs to the observation deck. Shoving open a heavy metal door, they cautiously walked outside with guns ready.

"When we got outside we saw five or six men handcuffed together. The men huddled together in a tight packet and even against the wind and rain they kept glancing up to a makeshift pole. They were chained to it and to one another."

"That's it," Allen said. "There's the bomb."

"It's suicide," Kelly said while looking at the television antenna looming up into the mist above them. The package was strapped near the top of the antenna.

"Besides the bomb we've got lightning."

The Lieutenant pointed. "That's it," he said. "The man says it's detonated electronically from a safe distance."

Kelly's portable phone rang. "Kelly."

"This is dispatch. We have a ransom demand."

There was a long pause and dispatch suddenly yelled out. "Get out of there!"

"Allen and I flung ourselves down the stairs shoving a patrol woman and another cop ahead of us. The explosion caught us at our backs, lifting us, burning us.

"A blast of hot gas and flame licked at us as we scrambled down three floors. A fireball bounced off the wall ahead of us. The fire continued down the stairwell. I heard people yelling further down.

"Allen had burns across the small of his neck and I along my back and arms. We sheltered the officers who were most unharmed except for their fall down the stairs."

"Kelly?" Mullins asked.

"More bad news?"

"We got to go up again."

"Say what?" Kelly stared at Mullins. "Go up? Why?"

"The ransom."

Mullins held out his portable telephone. "That's the ransom. We got to clean up what's left up there. All of it. We, you and me, have got to personally put the parts into body bags. Allan can stay down here."

"What?"

"That's the ransom. That's the demand?"

"Some bastard dares to demand a ransom after he's killed those people? Screw him," Kelly said.

"You talk to him." Sergeant Mullins handed Kelly the cell. "Yeah! This is Kelly!"

"Kelly?" Without waiting for a reply a gruff sounding man let his reasons be known. "Those winos are proof. That's all. Proof."

"Proof of what?"

"Of my credibility!" he shouted. "When I tell you something from now on you believe it!" He hung up. Kelly handed the phone back to Mullins.

The medics were the first onto a small section of the roof. In a few minutes they came back down the twisted partially blown out stairway.

"Sorry," one of them said, "No survivors," and he hesitated and added, "Better rig a fire hose to take up with you."

"There's still fire up there?" Kelly slowly got to his feet, looking up the blackened stairway. Kelly heard the wind push more rain down the darkened stairway.

From where he stood Kelly saw a big hole where the door leading to the roof used to be. Water dripped down his burned neck.

"Some fire here and there," Mullins said. Kelly wondered why they would need a fire hose as rain was beating down in buckets.

"Mullins and I made out way up to the roof. We saw why the hose was necessary. We'd be here all night putting body parts into plastic bags. Little bits and pieces were sloshing around in puddles."

As the rain kicked up it swept across the roof. The firemen had rigged portable lights. Damn," Kelly complained,

"That bastard picked a miserable night for this."

In the garish lighting he could barely make out various identifiable parts swirling around and heading for a drain.

"Damn," Kelly said again, "We need a hose."

In thirty minutes the fire department had a hose rigged to the one remaining functional standpipe. Firemen put out little hot spots along the roof.

They gave the hose over to the cops who hosed little bits and pieces into a common body bag. Five torsos were intact.

"The bum said six people," Kelly said. Only five had been lashed to the antenna. Kelly wondered if the bum had inadvertently counted himself twice.

"Another call," Mullen said. He walked out of the hole that had been the door. Kelly grabbed the phone.

"What!"

"Three hundred thousand. In small unmarked bills!"

"You already killed "me ya lousy bastard!"

"Look across Lafayette Street! Use your search beams! You'll see 'em on the roof! This batch you get back alive if you do as you're told!"

Kelly gave the phone to Mullins. Through the rain Kelly could barely make out some people huddled together on a roof far below them. They were hostages hanging in hammocks on the side of a building across the street. Kelly looked almost straight down on them.

"Can we identify them?" Kelly asked. Mullins didn't answer. Kelly saw the Fire Chief staring at him.

"Your wife and the Mayor."

"I wasn't sure what I was looking for. To the east is the ocean. To the west Lil Po Lake. To the north there was the Pewit building with my wife and the Mayor. I looked again at the roof. In the back of my mind there was something that the old bum had said that didn't add up.

"The bum said that six people were hostages, but we only counted five. The bum also said that the guy was on the roof, but we didn't see anybody else."

Mullins walked away. Kelly walked a few paces north, around the roof looking, but there was nothing but the twisted wall of steel and brick that had held against the blast. Heavy gusts of rain ripped across the exposed roof. Kelly's line of sight traveled along the safety railing.

It was damaged, except for one rail section with two steel cables extending over the roof's edge. What in the hell were those cables for? What were they attached to? Kelly walked to the railing and looked straight down. Using his flashlight against the glare of police lights two bright orbs reflected light up at Kelly.

A man stared up at him from the safety of a mountaineer's hammock. The bum had said six and that included the guy in the hammock. There had been six people on the roof, but only five bodies. Kelly knew that the killer was still on the roof; and, in fact was looking up at him from the mountaineer's hammock. The killer raised his hand out and pointed a small radio transmitter at him. "Back off," he yelled. "Or they'll die!"

"I could see it was a reverse pressure switch. Release of his thumb on the switch would detonate the bomb across the street. I looked around, but Mullins hadn't followed me over. I was alone with the son-of-a-bitch."

"Back off!" The bomber's voice was shrill.

Kelly took out his revolver in his right hand and used the flashlight with his left hand to illuminate the guy in the hammock.

In the hammock the man looked like a wet rat. Kelly aimed at the bomber's right knee. Kelly checked the roof one more time. As curtains of rain obscured the brick wall he looked again at the man forty feet below him. The man wiggled around in his hammock gesturing with the detonator by waving it over his head.

Kelly waited until a peel of thunder covered the bark of his .38 police Special. The bomber shrieked as the round tore through his right knee. Kelly grimaced. Damn. He could see the man's thumb was still on the trigger switch.

"Why?" the bomber yelled twisting around in the hammock and writhing in pain.

"Why?" Kelly shouted back, "To prove a point."

"Point? Point? What point!" the man shrieked.

Kelly slowly aimed the revolver at his left knee and squeezed off a round. Even at forty feet down Kelly saw the man's eyes roll back in his head.

"I let myself enjoy the moment. I mopped rain out of my eyes and grinned, but the bastard still had his thumb on the switch. He must've still been dreaming of collecting his ransom."

"Why? He shrieked."

"Because," I shouted back, "to prove my credibility. The city is broke. We don't pay ransoms!"

Two sharp cracks of thunder echoed Kelly's two rounds piercing the steel cables holding the extortionist's hammock to the building. Kelly watched as the hammock plummeted into low clouds far below. The bomber disappeared into a deep gathering mist.

"Somewhere along the way down he released the switch."

Jacob listened. That a cop like Kelly would facilitate the death of his wife by forcing a bomber to kill her in one way stunned him; yet, in another way it seemed natural to the world. People *were* violent. Yet, mostly there always seemed to be something behind it, something to be gained or some real or imaged wrong to be made right.

"Why did you do it?"

Kelly made himself comfortable on his sleeping bag. Jacob was getting used to seeing men in the buff. Already with less than a day, in a house turned into a prison, he was adapting. Jacob swallowed hard, because he didn't want to adapt too much.

"I was a good cop for twenty years," Kelly said. "I mean an honest cop. Believe me, Ebbtide, there aren't many of us left." Kelly propped his head up on his upturned hand, watching Jacob.

"Yeah I wanted her dead."

"Why?"

"She was screwing the Mayor."

Jacob rolled over. "Look, Kelly," he said trying to reason with him, "Divorce or bad faith aren't reasons for murder."

"Maybe," Kelly mumbled. "If they were only lovers," he said, "I could live with that, but they chose to involve me."

"Involve you?"

"The Major was running a racket. Counterfeiting gasoline ration coupons. Sandy, my wife, was in on it. I knew she was doing something, but didn't look too hard."

"How did they involve you?"

"The law was catching onto them. They set me up. They planted evidence against me."

"Ooh," Jacob breathed. That kind of setup he could understand, although he doubted that, when push came to shove, that he would ever kill Gladys.

"And the bomber?"

"What about him?" Kelly asked.

"The whole thing was an accident? Come on," Jacob said. "Your wife and the Major accidentally happened to get kidnapped?"

"Naw. Them bastard kids. Those goddamn gangs," Kelly said. "Every city has them you know. Boston, New York, Washington. All I had to do was pick up a couple of them.

"The leaders I mean. I kicked the crap out of them. I let them overhear a make-believe conversation on my cell phone.

"I knew the information would get back to the gangs doing the kidnapping. I mean its big business. I made stupid you know? Blabbing, bragging, how much money my wife and the Mayor had socked away by selling bogus gasoline coupons."

"The kids took the bait?"

"Sure."

"And the guy in the hammock?"

"He was the engineer," Kelly said. "Ebbtide, you're a scientist. How many people can actually rig a bomb?"

"Not very many," Jacob admitted.

"So the mountaineer was in for a percentage?"

"Probably, although dealing with those kids is risky. The little brats are thieves at heart. They needed the engineer to pull the job and they cut him in for a piece of the action."

Jacob watched as Kelly bunched up a part of his sleeping bag as a pillow and was asleep almost immediately. Jacob sat for a while looking around the quiet room.

The lighting was down. Up on the wall a video camera's red light blinked on and off reminding Jacob of American satellites over Tel Aviv.

8

Deep in the night Jacob woke up. Where was he? Half asleep, he turned around and lay on his back with his feet out of the sleeping bag. Overhead a fan whirled. Rubbing his face, trying to wake up, he tried to remember what had happened and where he was. He got up and stumbled around in the semi-darkness.

In an instant he was on his feet prancing on a carpet. As he shouted lights came on around him and others jumped up shouting. When he yelled they yelled. When he screamed they screamed.

Then a big man next to him rushed out while Jacob was prancing up and down and yanked him in the direction of his sleeping bag.

Jacob collapsed face down, his feet kicking about madly in the air over the sleeping bag. Someone doused water on his badly burned feet.

"Ice!" the man shouted while throwing warm water on Jacob's blistered feet.

"Shay," the man shouted, "Pull that bookcase over."

The other man came over, looked down at Jacob and then moved the bookcase. Then Jacob's feet were up on the case. Jacob was on his back, lying there, crying hysterically. The air smelled of burned flesh.

"We need help!" the man shouted as the door to the loft opened. Two other men ran in. They knelt down by Jacob with a medical bag. His blistered feet radiated pain all the way up his leg causing his leg muscles to cramp.

The men with medical bags put salve on his feet and lightly bandaged them. They gave Jacob a sedative. Within two hours the pain mercifully subsided and he fell into an exhausted sleep.

The following morning Gladys came to see him. Jacob had spent the remainder of the night flat on his back with his feet supported in the air by the bookcase.

"Well, Sir," Gladys said, standing outside his area, "You've had a bad experience."

Jacob groaned. He put his arm over his eyes, but peeked out from under his arm to see her.

"Are you sedated?"

"Yes," Jacob mumbled.

"Your first hotfoot is free," Gladys said. "We won't add anything to your time. You'll be laid up for a couple weeks. You can have the radio."

"It was an accident."

"We know Jacob. I have the medics report."

She paced around looking at him. "You must understand, Jacob, that any more mistakes will cost you more pain plus six months. Understand?"

"Yes," he sobbed feeling sorry for himself.

"Please don't cry, Sir. It's unbecoming for a man. Don't you think?"

Jacob swallowed air and moved his feet around as he tried to get comfortable. He muffled his sniffles and said nothing.

"We have the grid on low power for you Jacob. It was inevitable that you would step out. Everyone does. Absolutely every new prisoner gets a hotfoot. They all do."

"My feet?" he questioned.

"Your feet will heal. Second degree burns. Little muscle damage, but you stayed on the carpet dancing as I understand it. I won't ask what possessed you to do a jig. Were you being funny?"

"God damn you!" Jacob shouted.

He heard Gladys take a deep breath disgusted with him. "Two weeks Jacob. The snails will wait for you."

She pulled a small cell phone out of her purse, tossed it onto the countertop and turned to Car Wreck. "Route all his calls through your station and monitor and keep tracks of all charges. He gets one call a week now. Five minutes."

"Yes Judge," Car Wreck said.

"By the way Jacob I've got word from President Leviman."

"That bastard?" Jacob asked. "What's he want?"

"He was interested to know that you're in the United Remnant."

"Does he know I'm in jail?"

"No," Gladys said. "I have a confidentiality agreement with the government. We have you."

"Does Leviman know that?"

"He thinks you are working for the Americans. He thinks you always had some deal cooked up with them."

"But I didn't!" Jacob shouted. "Everybody thinks I love deals!"

"My god," Gladys said. "You really don't know what kind of man you are. Do you?"

She walked toward the door. "Oh one more thing, Jacob," she said. "President Leviman thinks that our daughter is here in the United Remnant."

"Is she?" Jacob asked.

"I'm asking you!" Gladys shouted. "Did you have something to do with her disappearance?"

"No!" Jacob shouted. "If you find out anything Gladys you must tell me!"

"I don't have to do anything, Jacob."

"You must tell me," he insisted. "I'm worried about her!"

When Gladys left Jacob looked around, rolling over to his side and putting one burned foot on top of the other. He pulled himself up on an elbow, craning his head around.

The other prisoners were out working. Car Wreck was at the computer, head down, paying no attention. Maria stood up.

She came around the computer station with a bottle of water. "Drink lots of water Jacob," she said handing him the bottle.

"You'll heal faster."

Jacob gratefully accepted the bottle. He took a long swig, but then realized he needed to urinate badly. His feet were propped up on the bookcase.

"Maria," he said handing the bottle up to her,
"I need to pass water." Maria looked down at him
indifferently.
"I need to pass water," Jacob said again.
She pointed over to the toilet at the back of his area. Jacob
shook his head and pointed to his burned feet resting up on
the bookcase.
"Drink," she said giving him the bottle again. She left his
area, but when she got back to the computer she added,
"Then use the bottle."
After finishing the bottled water Jacob realized how
dehydrated he was, but when he positioned the empty bottle
he realized he would be peeing uphill.
"Help," he called plaintively. "I need help." This time Car
Wreck came into his area.
"Can you help me?" Jacob asked.
"No," she said firmly. "Are you a baby?"
Jacob put the water bottle down. "It won't work like that. I
need to get up on the toilet. Please," he begged. Car Wreck
turned away and walked to her station. When she got back to
the computer workstation she looked at him.
"If we do that for you, Ebbtide, our lives would be hell! See?
These others would have us cleaning up after them."
"Please," Jacob begged.
In was no use. Car Wreck went back to work, but then
Jacob watched Maria get up and whisper something to Car
Wreck. Jacob swung his legs of the bookcase, resting his heels
on the floor. They felt comfortable. Rolling onto his stomach
he crawled on his hands and knees over to the toilet and then
pushed up on it. The toilet was too high. Somehow he had to
turn around, without putting weight on his feet and get on
the toilet.
He tried to figure out how it could be done. He struggled
with the physics of it for a few minutes and then yelled,
"Help!"
A towel suddenly hit him in the head.
"Use it as a diaper," Car Wreck commanded,

"But remember that's your shower towel!"

Jacob knelt in front of the potty with his head half down in the bowl. His bladder was bursting, but he forced himself to think. "I'm a scientist," he told himself. "This is a case of weight distribution."

He took hold of the toilet and slowly rose, tummy first, up onto the pot. While lying on it he looked hard at the anchors. The toilet was bolted to the floor. It wouldn't tip.

"Look," Maria pointed to him, "That one is smart. He's figuring it out in a few minutes."

"Sooner or later they all do," Car Wreck said.

"Most pee on the floor first."

"One peed on the strip."

"Yeah," giggled Maria. "The hotfoot went up his stream."

"He cried," Car Wreck said. "He let out a whoop." She laughed out loud. "I saw the spark."

"They cry," Maria added. "They cry like babies."

"Men are babies," Car Wreck said.

Then, with the towel rolled up at the base of the toilet Jacob kept the balls of his feet up off the floor and for an instant he thought he saw Maria give him the high sign. With effort he pivoted around and got settled.

The following day Car Wreck tossed the cell phone to him.

"Ebbtide."

"Who?" Jacob asked.

"This is Handy."

"How did you get through?" Jacob asked.

"Dialed the number," Handy said. "Wasn't difficult."

"Can you help me?" Jacob asked.

"Maybe," Handy said, "although I don't know what I can do."

"I need help," Jacob said. "I'm in prison," Jacob said. "I want to work for the government, but first I need…"

"Won't they let you out?" Handy asked. "What kind of a deal did you make with those bastards?"

"Who?"

"The Kamsi."

"What?" Jacob asked.

"You tell me," Handy said.

"I'm in the Hines Prison House. I'm right here in New Boston. I need to contact the government. They need my technical assistance."

"You're a liar," Handy said. "You were a smart little spy for Israel years ago and you haven't changed one damn little bit."

"But I'm not lying!" Jacob yelled. He looked up to see Gladys walking into the loft. She stood in the doorway looking from him to Car Wreck.

Jacob's blood pressure spiked when he saw Gladys put a hand to her right ear. She had an earphone. She was listening to his conversation.

"I'm not lying," Jacob said again. "I need help."

"According to everything I can find out," Handy said, "You're working for the Kamsi. Even the return phone number is in the Kingdom. If you want out you'll have to get out on your own. Frankly, I hope they fry your Jewish ass."

Jacob looked to see Gladys give Car Wreck a sign. The call was instantly cut short. Only a dial tone rang in Jacob's ears.

"Put that name and number on the delete list," Gladys said.

"You can't do this!" Jacob yelled. "That man believes that I'm in the Kingdom!"

"Is that right?"

Car Wreck came over and collected the portable phone. Jacob knew that Gladys had covered all of the bases. She knew the system and she had closed off his last remaining hope. Jacob hoped that he could make contact with somebody on the outside and that they would intercede with Gladys. Hope was all he had left.

*　　*

From her point of view Gladys considered the whole situation rosy. She spoke with Leviman, but never admitted to having Jacob in her jail. She did admit to talking to Talya about once a week.

Leviman had told her that Talya was in the Remnant with two of Jacob's researchers, Herman Burger and Emil Zada, and that they were someplace safe, although he didn't know where. Apparently, he said, they thought they were in trouble, but they weren't.

For herself, Gladys didn't want any part of Middle Eastern politics and simply avoided the whole thing. She did speak with Talya almost every week, but even her own daughter wouldn't tell where she was. Gladys didn't like being kept in the dark. She was a Judge and she still couldn't find out where Talya was hiding.

"Where is she?" Gladys asked Leviman.

"You talk to her. Don't you?"

"She won't talk about it," Gladys admitted. "Will you tell me?"

"I don't know," Leviman said. "We've bribed all the usual sources and threatened the others," but nobody knew the whereabouts of the missing trio.

"You communicate with her?" Leviman asked.

"Of course, but she's tight lipped."

"It makes me mad," Leviman said. "What about Jacob?"

As for Jacob, from Leviman on down through the operatives in the Mossad, it was agreed that Doctor Ebbtide, as head of the research department (and who had *hired* Dexter as manager of the Water Plant) should have been more diligent.

Israeli intelligence had traced Jacob's movements to the Remnant had then had lost him. He might be in the Remnant or back in the Kingdom where his knowledge of genetics would be put to good use.

"Any idea where Ebbtide is?" Leviman had asked her.

"Not a clue," Gladys said. "I'd like him to come back here so I could jail his ass for kidnapping my daughter."

"We need to find him," Leviman said. "We need him."

"Or kill him?" Gladys asked.

"If we wanted dead he'd be dead!" Leviman shouted.

"I shipped him to the Kingdom," Leviman said, "because I wanted to test him!"

"Test him?" Gladys was incredulous. "Test him for what?"

"His loyalty!" Leviman said. "He hired the people at the Water Filtration Plant and others. All of them turned out to be traitors! I didn't want to believe that about Ebbtide."

"Jacob? A traitor?" Gladys asked. "Are you drunk? He's the most dedicated Jew I've ever met. I should know. I married him!"

"All he had to do was come back!" Leviman shouted. "We are working, more or less, with people inside the Kingdom. A Sheik by the name of Kasan is somebody we trust. We sent Jacob to him with the expectation that he would come back!"

"Are you crazy?" Gladys shouted. "Jacob probably thought you were going to kill him!"

"Is he stupid?" Leviman shouted. "If we were going to kill him why in hell would we turn him over to the bastards in the Kamsi Kingdom? My God!" Leviman bellowed.

Gladys saw Leviman's predicament. To the President of Israel it was a simple affair. Cut Jacob loose and see if he swam back into Leviman's waiting arms. It was, to Leviman, a simple test of loyalty.

"You really don't understand Jacob do you?" she asked. "To a man like him any rebuff, any slight, is a blight on his honor. He wouldn't see your tactic as a test. In fact it would only confuse him. The poor bastard really doesn't know which way is up!"

"We've got to find him!" Leviman shouted. "He won't survive in the Kingdom. Once they've pulled the information out him they'll kill him! Things are awful over there. "Even Muslims are getting their heads chopped off for minor infractions like spitting on the sidewalk."

"Jacob can take care of himself," Gladys said. "We've got to find him! If he contacts you just tell him to come home! Please. Will you do that?"

"He won't contact me," Gladys said. "I'd be the last person in the world he wants to see." She hung up.

"My," she remarked to one of her staff at the Courthouse, "Long talks make me hungry."

An hour later the Judge walked out of the courthouse to a public phone and dialed a number.

"Hello?" a woman answered.

"Hi," Gladys said. "I just called to ask after you."

"I'm fine," Talya answered. "But I'm worried about Daddy."

"Don't worry about Jacob," Gladys said. "He's a big boy. He can take care of himself."

"What if something happens to him?"

"I talked to President Leviman a few minutes ago," Gladys said. "I'm telling you what I told him. Don't worry about Jacob."

'God,' Jacob thought as he lay in his sleeping bag. Now what are my choices? The Kingdom, Israel (probably a firing squad) or stay as a jailbird in the United Remnant? While his feet healed he spent many hours thinking over his situation.

Two weeks passed. Life is a routine. Even in prison. Two weeks to the day after his accidental 'hotfoot' Jacob slipped into a pair of socks and shoes and joined his place on the chain gang.

Once, out in the yard, pushing a hand mower, he caught a glimpse of Gladys standing at a window watching him. A bile of outrage rose in his throat. He had been mocked, burned, and turned into a slave, for which he would have to pay *her* restitution!

"Damnit," he cussed to himself, "I *will* find a way out of here!"

He had to. Five years of this would kill him. Maybe not his body, but his spirit would be broken. As he was thinking he pushed the lawn mower.

He shoved on the mower and bent over, with Gladys watching from the window and put all his strength into it. It was one tough machine to push.

Once or twice he stopped to mop sweat off his brow and knelt down to look at the mower. Something was definitely wrong. It took all his strength to get it to move. He shoved the damn thing with his tired legs and arms trembling with exertion.

Was he a baby as Car Wreck insinuated? Had he lost all strength? He dug in his heels, his arms stiff, shoving with all his might to budge the lawn mower.

"What's with those mowers?" he asked Kelly that night.

"Brakes," Kelly said. "The Judge adjusts each machine herself. They can roll easy or hard. It all depends what kind of mood she's in."

"Brakes?" Jacob asked incredulous. "Brakes? On a god damn lawn mower?" Jacob raised his voice. "Brakes!"

Car Wreck and Maria looked over threateningly.

"Watch it," Kelly advised. "It's easy to gain time around here. Infractions mean time. Don't forget that."

"Yeah but," Jacob's mind boggled with the idea of putting brakes on lawn mowers.

"Yeah but, Kelly, I mean, brakes on a lawn mower? My god! What kind of bastard does that?"

Kelly leaned over whispering, "I think you mean bitch. Not bastard. Bitch. Got it? And you should know. You married her."

Sometimes in deep summer when Jacob was mowing one of the huge rear yards, the Judge came out of the house and sat under a wide sheltering beach umbrella sipping lemonade. She watched him work. She did not speak with him nor offer any refreshment. At night after ten hours of hard physical work Jacob's knees wobbled. His muscles ached, but more than the physical pain was the realization that she had set the resistance on the lawn mower a notch higher.

Her adding injury to the insult of being locked up aggravated the hell out of him. A lock on the axle controlled the brakes. There was not a damn thing he could do about it. He was weak with exhaustion, but he was still a scientist. He exercised his brain every day trying to figure a way out.

The season turned to late summer in New Boston. Jacob toiled in the heat. Sometimes he picked snails out of her garden, but mostly he mowed. Even in the heat he forgot to wear a hat and even with the mower on the highest brake resistance setting, with hard practice, he pushed easily with one hand. He was, after months of hard work, in the best shape of his life. He was accustomed to the heat and to hard physical work.

As the summer drew to a close he actually felt a sense of pride. Moreover, he lost weight and become brown with the sun and then finally, after weeks of puzzling out the problem of how to escape, he knew the answer.

"I have a way out," Jacob confided to Kelly late one night. "You interested?"

"Damn straight." He scrunched over listening.

"I've been assigned to Stonegate."

Jacob felt Kelly's fear radiating from him. In the dim light of the barracks Kelly's eyes also betrayed his fright at ending his days in the middle of a radioactive waste dump.

"I was double-crossed," Kelly said. "I paid good money to rig the system to get me in here, but when my money ran out the bastards screwed me."

"I know the feeling," Jacob said.

"You have a plan?"

"It'll cost money," Jacob said, "but I can get money. We need to get word out. Can you arrange that?" He rolled over putting his hands under his head and looked over. Kelly thought for a moment.

"Getting word out is no problem, but it'll cost a couple thousand gold dinars."

"I can get it." Jacob thought for a moment. "We will need a go-between," Jacob said. "Someone that can come in here, talk to us, take messages and leave and not be noticed."

"Huh," Kelly said and Jacob saw the fear disappear from his face. "*That* is not a problem." Jacob followed his gaze over to the computer console.

"Them?" Jacob asked.

"They are trustees," Kelly replied.

"Trustees?" Jacob asked. "Aren't they almost cops?"

"That doesn't make them honest."

"What does it make them?"

"Smart crooks."

Jacob shook hands with Kelly. It took two more weeks for the plans to be set. At night, after the trustees had left for the day he and Kelly rehearsed the plan to each other.

Car Wreck and Maria were only involved in the part of the plan that involved getting word out to Sheik Kasan that Jacob had a change of heart.

Kelly made to clear to Jacob that the trustees shouldn't be involved in any of the actual escape plans, because part of their sentences would be cancelled if they snitched on other inmates—the Car Wreck and Maria could only trusted so far.

Then, on a day like any other day Jacob walked out of the shower room with a slopping wet towel around his waist. When he got to the computer station he stopped momentarily looking around.

The other prisoners were in their areas sacking down for the night. Jacob stood near the console. On his far left Justin was listening to his portable radio. Shay was shuffling some magazines, D'Lill was already asleep and Kelly was sitting, buck naked, in his stall watching him.

"Move it!" Car Wreck commanded.

Jacob took a step toward the now deactivated carpet, swung the wet towel off his waist, twisted it up and then lunged for Car Wreck. Jacob caught Car Wreck off guard. Wrapping the sloppy wet towel around her head and shoved the towel into her mouth.

Car Wreck reared back bringing her big beefy arms up defensively. Kelly rushed to Jacob's area and then jumped across the carpet lunging at Maria.

As Car Wreck tumbled out of her chair an alarm sounded. Jacob was astonished at how fast she was. Her beefy arms were filled with muscle. Her bones were tough as steel.

"God!" Jacob breathed. No wonder she won East German beauty contest! Car Wreck was on her feet almost instantly. Her meaty fists slammed into Jacob's nose like a beer truck.

She spun around and bent down reaching with a rough meaty hands grabbing Jacob by his short parts and yanked. Jacob howled, dropped the towel and collapsed face down toward the floor where his nose made hard contact with Car Wreck's bony knee.

When Jacob woke up his face was bloody and his nose was broken. He lay on the floor behind the computer workstation moaning. He saw Maria sitting in her chair at the computer station with Kelly at her feet unconscious. Medics were working on him.

Fifteen minutes later Jacob's nose had a large bandage. Car Wreck dragged him out from under her workstation to his area where he was handcuffed.

Kelly was awake and mumbling tearfully. "I never should've listened to you," he sobbed. Jacob managed to get his eyes open wide enough to see his partner in crime flat on his back also handcuffed sobbing tearfully.

"Whose idea was this?"

Jacob looked up to see Gladys. She stood over Kelly with her arms folded tightly against her breasts. Kelly was in a mood to be contrite.

"His!" he sobbed pointing over to Jacob. "He said he had a plan. Ah, god, what's going to happen to me now?"

Jacob was groggy from the beating he had taken from Car Wreck and heavy treatment from the medics. His nose felt like it weighed two pounds and that he'd never be able to breathe through it again. He gulped air through his mouth.

"My plan?" Jacob blubbered. "Whayaya mean *my plan?*"

"Yours!" Kelly shouted. "I never should have listened to you!"

Gladys walked over to Jacob. "Well, Sir," Gladys said emphatically. "Now you've done it. You attacked the trustees."

She paced the floor in front of Jacob's area for a moment, shaking her head. Going to the computer workstation she pulled the clipboard off the workstation.

"It's all here Jacob," Gladys said. "It's official. I can't help you."

"Please," Jacob begged. "I don't know what I was thinking…"

"Don't," Gladys said, but Jacob broke down, crawling to the edge of the carpet on his knees with his swollen nose close to the floor, head over, in abject subjugation.

"Please Judge," he whined. "One more chance. I went crazy. I couldn't take it anymore. Lawn mowers! Lawn mowers with brakes! Brakes! Puullease!"

"Shut up!" Gladys commanded.

She wrinkled her nose in disgust. "God Jacob," she said hurriedly, "I never thought you'd break like this! I remember you running a con on that bitch Jennifer. You ran with Frank O'Hannon. God!" she cussed. "You were a Jew with guts! I never thought a lawn mower could bring you down like this!"

"Please please," Jacob begged groveling at her feet. "Please, Judge," he sobbed, "Don't send me away!"

Next to him, also on his knees, Kelly was muttering slavish apologies. His forehead was pressed to the floor; hands handcuffed together pleading for mercy. "Jacob made me do it. It was Jacob! Jacob!"

Gladys stared at the two men, now elbowing one another out of the way as they each tried to kiss her shoes. The Judge jumped back as she saw them on their bellies crawling forward. Gladys dropped the clipboard on the work desk and hurried to the door. When the door slammed shut behind her Jacob stayed on his knees, face to the floor, as Car Wreck came up standing over him.

Kelly brought his head up off the floor and scooted back quickly as Car Wreck bent down and grabbed Jacob by his ears. She lifted his face to her. When Jacob came up his eyes were filled with tears. He blinked. The fat bitch wasn't wearing any panties. She reached down and shoved his face up between her fat hairy legs.

"Take a good look!" she shouted. "You little Weasel! Where you're going my cunt will look like heaven!"

She released Jacob's head. He bounced his wounded beak off the floor, rolled over, and cried out loud.

The following morning all inmates were quiet and sulking. On days of prisoner transfer there were no work details. All inmates were confined to their areas.

The same cops that had delivered Jacob to Gladys fitted the shackles on his wrists and ankles. He winced as metal clicked.

Car Wreck put shackles on Kelly. She seemed to enjoy her work. Kelly was dogfaces. He hadn't used his electric razor or combed his hair.

Car Wreck finished getting him shackled and then reached over and zipped up his new orange prison jumpsuit. She surveyed her handiwork up and down and looked to Jacob.

"Smuck," she said. She pointed up to the video cameras. "Smuck," she repeated. "You couldn't have made it."

Jacob looked to the cameras and then in her direction, issued a loud sob and sniffed his nose. A big fat salty tear trickled down his cheek.

"You should never have listened to me!" he scolded Kelly. "If it wasn't for you I wouldn't have tried to escape!"

"Your big plan!" Kelly yelled, "Ha!"

"Shut up!" Car Wreck shouted. "Both of you!"

Jacob wasn't sure, but thought the same van that delivered him to Judge Hines was taking him to Stonegate. He sat next to Kelly that peppered the cops with a running monologue on how Jacob had conned him into the plan for a jailbreak.

The van stopped about three miles from Stonegate and Jacob and Kelly were transferred to a dilapidated helicopter with no doors for the passenger section. The doors had been removed.

"Get into these!" a pilot said as he handed two black cloth hoods to the inmates. With effort, because of the short chain linking his wrists and ankles, Jacob squeezed his head into the black cloth.

With assistance Jacob stumbled into the helicopter and sat very still as the pilot strapped him into the seat. After a few moments the engine started. The helicopter shook violently, bounced around for a few seconds and then forced its way into the air.

The ride was short and Jacob knew even under the black hood they were at the prison by the odors of hay, grass and putrid pond water.

The helicopter banked in a steep turn. Jacob felt dizzy. The pilot reduced power. The machine settled down to a rough landing.

"Get them sacks off!" a crewman ordered.

Jacob blinked as the cloth came off his head. He saw that they had arrived at a collection of trailers, hovels, huts and holes-in-the-wall that collectively comprised Stonegate penitentiary.

In the distance, over mounds of debris encircling the prison, Jacob saw other helicopters. Two ships followed one another at low altitude toward the City.

Lowering his sight he saw a collection of small metal trailers on a patch of dirt surrounded by a vast wasteland of twisted metal heaped up into rambling mounds of radioactive waste. Heat shimmered off metallic piles in waves of undulating shimmering air.

Jacob stood with Kelly beside the helicopter looking around to see that the helicopter had landed about one hundred feet from a ramshackle collection of dilapidated trailers. Once off the helicopter a cop grabbed the hoods.

"Hurry it up!" the pilot shouted over the rotor noise as a kid came up to the helicopter. He had the word 'trustee' roughly blocked out in white letters on his stripped prison shirt.

"Sign for them. Hurry it up!" the pilot shouted again.

The transfer process was always the same. A clipboard came up, somebody signed for the merchandise and the transfer was complete in a few seconds.

When the trustee signed for the new prisoners the pilots jumped back in the helicopter and everybody ducked as rotors kicked up loose grass and dirt.

"New fish this way!" the trustee shouted over engine noise.

They followed the kid into one of the trailers. They shuffled along behind the trustee like two very old men into a tiny trailer filled with steel chairs meant for little kids. They were in an old schoolroom trailer. Dirty childish pictures still adored the walls.

Maps tacked up on one wall showed what the United States had once looked like. A small desk up front and a dozen student desks, far too small for adult men, rounded out the furniture. A man sat behind the teacher's desk.

"Is this it?" he asked the trustee.

"Yes, Warden," the kid said. "These two are the last for the day."

The Warden stood up. He wore a brown suit, yellow shirt a tie with happy rabbits. Around his waist, Jacob saw a heavy contraption. It looked like a lead diaper. The diaper was suspended around the middle by a collection of heavy rubber suspenders draped over his shoulders and he hung his thumbs in the suspenders as he smiled in greeting.

At the trailer door two guards entered and stood to one side as the man in the suit went around to the front of the teacher's desk and faced the room. Jacob saw that they also wore the metallic chastity belts, but unlike the gentleman in the brown suit they wore dirty overalls.

It wasn't only their clothes that were dirty. The guards looked as if they hadn't washed in months. In the stifling air ripe body odor whiffed into his swollen nostrils.

The man in the suit cleared his throat. He sat down on the edge of the desk with his lead diaper climbing up his butt.

"Welcome," he said. "My name is Doctor Donald Veeby. I am warden of this facility. I make it a point to welcome all of our new arrivals."

Jacob watched the trustee's fan out around the small room, taking up positions along the sidewalls. One fidgeted around, his fingers digging deep into his lead diaper trying to scratch in a delicate place.

"Right off the bat," Veeby said, "I want to dispel notions about Stonegate being a pesthole of lingering death."

He winked at a guard. "Why I don't think we've lost more than thirty percent of our inmate population to radiation poisoning over the past year or so."

He paused for effect. The trustees giggled. One of them in the rear of the room used both hands to dig into his diaper to scratch a serious itch deep down.

Jacob fugitively watched the other new inmates sitting in their seats. They looked completely subdued. He counted nine 'new fish.'

There was silence in the trailer. Only a distant drone of departing helicopters reverberated in quiet air. Around his right ear a fly bussed then, as if realizing where it was, it flew out an open window.

"Seriously folks," Veeby said raising his arms in welcome, "I can't tell you how happy I am that you've decided to make a career move to Stonegate."

Veeby walked around the desk and opened the top desk drawer. Pulling some vanilla folders out, he tossed them to the desk. He stacked them up and then walked around to the front of the desk again.

"While we like to joke about radiation misconceptions abound. Some things are true. Others not." In the back of the room a guard grunted.

Jacob pivoted in his chair to see the guard drop his hands to his pants, locking his thumbs in a heavy leather harness covering his lead diaper.

The Warden walked to a blackboard, picked up chalk, and drew a box for the prison and a bunch of lines for the streams, ponds and mountains of debris.

Through the middle Veeby stretched a straight line cutting across the upper edge of the prison/dump complex. "That's the railroad bringing debris into the facility," he explained.

"Now, some of you are aware of the situation. We simply had to find some place to stash nuclear waste; and well, you know, every time the Government came up with a billion dollar solution some environmentalist screamed bloody murder. So," he gestured outside, "When the Government went bust and crime became rampant the solution became obvious." He raised his arms expansively.

"Why not use radioactive waste to control waste like you?"

Veeby nodded as if agreeing with himself, he added,

"Good. You understand. You had to be flown in here and the only ways out are by air or in a lead coffin."

In the back of the room a hand went up. "Sir?"

"Yes?" Veeby asked.

"I was in nuclear medicine in the Army, Sir. I know what this stuff does." The prisoner put his hands around his waist, "Do we get the special panties?"

"No son," Veeby replied in a fatherly voice. "You are new fish. Radiation affects sexual activity first. Men become impotent. And," here Veeby's voice became that of lecturing professor, "Aggression and sexuality are linked. Reducing your libido works to our advantage. New prisoners stay close to the outside. Hardened felons, rapists, murderers and others get deeper digs, but only if they obey the rules. They get to dig their own burrows in soft, but radiation free earth if they don't cause trouble. Otherwise they are put right out on the perimeter. They always simmer down," he chuckled, "Or light up whichever comes first. When they are ready to behave we provide shovels."

Here and there prisoners crossed their legs and some put their hands over in a fig leaf protective gesture.

"Is there some way to transfer out?" another man asked.

"Not that I know of," Veeby said, "but I've only been here a short while." He gestured around the trailer. "Even the metal in these walls were made from recycled cars and trucks. Of course," he smiled again, "We test it. Some of this stuff is hardly radioactive at all," and he raised his hands up into the air, his voice resonating in mock pleasure, "but at night, on film, this place lights up like a Christmas tree!"

Men fidgeted around Jacob, sliding their tiny chairs in together closer and getting distance between themselves and the walls. Veeby opened the trailer door as if preparing to leave, but stopped and gave a quick wink at a guard.

"Tough guys spend a lot of time tending grass out around the ponds. Do I make it clear? So be a good citizen and dig a deep hole."

While standing in the trailer doorway Veeby motioned for a trustee to get him those folders he had left on the desk. When the Trustee handed the stack to Veeby he picked through them and lifted one out of the stack.

"Which one of you is Doctor Jacob Ebbtide?"

Jacob raised his hand politely. "Is there a problem?" Jacob asked.

"A doctor no less," Veeby said, "I myself have a doctorate in chemistry." He seemed to wait for some response from Jacob who sat politely, but said nothing.

"Your file has a letter addressed directly to me by Judge Hines," Veeby said. "She strongly recommends preventative measures with you," his voice hardened.

"We don't tolerate trouble-makers here!"

"No trouble," Jacob replied submissively. "Honest."

Veeby read down through a document in the file. He tossed the file back to the trustee.

"Assign this man work. Let him mow the grass around sectors six and seven." Jacob watched the blood flow out of the trustee's face.

"Six and seven?" the trustee asked. "But…"

"Mow grass!" Jacob shouted. "I've done nothing but mow grass since I've been incarcerated!"

Around the room the other prisoners hung their heads. It was obvious to Jacob that the idea of living close to radioactive materials terrorized them.

"I've done nothing but mow grass!" Jacob repeated, but instantly regretted his outburst. He resolved to restrain himself.

"More of the same!" Veeby retorted.

Jacob wondered, but didn't ask if the lawn mower had brakes. He reasoned that there was no point in giving these people any ideas.

"Sectors six and seven," Veeby said to a trustee.

"Way out there Sir?" the Trustee asked.

"Do I have to repeat my instruction?"

"Right now?"

"Yes now!" Veeby shouted. The Trustee gestured Jacob out of his chair.

"Come on! Damn troublemaker."

"I want all of it mowed!" the Warden said as he walked away.

Once outside Jacob and the trustee went to a wooden box alongside the structure. The trustee pulled out a white radiation suit. There was only one radiation suit in the toolbox. Jacob stood as if paralyzed as the Trustee put on the suit.

"You can't do this!" Jacob insisted realizing that there were only one protection suit and two of them. The trustee grabbed Jacob by an arm and led him away from the trailer.

"This way Fish," the trustee said through an intercom in his protection suit. They walked out past mounds of scrap metal with the radiation meter on the trustee's belt beeping incessantly. They walked through one crooked lane and another and finally came to a huge grassy area about the size of two football fields.

Far to his right Jacob made out a raised embankment for a railroad track and then beyond that more metal that seemed stacked in every direction.

"Start mowing!" the trustee shouted. "There's the lawn mower!"

Jacob squinted, searching the rolling hills and saw a lone lawn mower standing abandoned near the crest of a grassy dune. Jacob and the trustee walked to the lawn mower and Jacob turned around to view his surroundings.

From his vantage point on the top of the hill Jacob saw a vast rambling collection of buildings. In the distance, on the embankment, a steam locomotive puffed into Stonegate. Dozens and dozens of boxcars and flatcars rolled in behind an old steam locomotive. Cranes and other equipment went to work lifting scrap metal off rail flatcars and dropping the scrap onto trucks that carried it away from the rails. Jacob stood for a long moment looking at the whole environment. He was scared, but his scientist mind still wanted to understand how the place worked

"Is that really a steam locomotive?" Jacob asked him.

"What's it to you?"

The trustee pointed to the lawn mower and pointed then quickly pointed to a dead rabbit on the ground. Jacob looked, saw the dead rabbit and over along the weeds two dead birds.

"Don't touch anything," the trustee said.

He brought his gloved hands up to Jacob's face.

"Too hot even with gloves. Got it?"

"Yes." Jacob said.

With a heavy heart he realized he was standing on ground with dead animals that used to be alive in the beautiful city of Boston. He felt sorry for the City and pity for himself. He looked around looked at the world and still felt wonderment at life. He was man enough to see beyond his own misfortune.

Above the air filled with dust from the constant piling of debris onto the heaps of radioactive junk. Jacob saw the sun shining brightly through the haze. The country was reverting to burning coal. Now early in the morning the air was filled with ash and soot. Jacob looked down at the lawn mower.

It the same model he had used at the Judge's. Jacob pushed the mower one way and then another making no effort to mow efficiently. He mowed around dead animals.

Here and there he shouted at whitetail deer that had wandered into the area, but it did no good as they were obviously tame. Jacob tried to run a deer off with the lawn mower, but it walked a few feet away and nibbled grass. Jacob mowed grassy hills and made good time. He was in good physical condition and the mower was the same 1920's model so popular before gasoline lawn mowers came into widespread use. After a hard day's work moving Jacob heard the Trustee hailing him from a distance.

"Get in here!" he shouted.

Jacob walked toward him pushing the lawn mower in front.

"Leave it!" the trustee shouted. "You a smart ass?"

"What?" Jacob asked.

"That thing is poison!" the trustee shouted.

"Never push anything my direction! Understand?"

"Yes sir," Jacob said politely.

Jacob followed the kid to a low earthen dugout reminiscent of film footage he had seen about World War I. Old railroad ties buttressed a low ceiling and shored up earthen sides. Earth cover sloped up away from the entrance. Far above, on the hill, Jacob saw trees growing and more deer grazing.

Once inside, however, the dugout opened to a brightly lit interior with yellow tiles on the walls and ceiling. A huge carpet covered the center of the floor. The trustee, still in his radiation suit, led him though.

"Induction heating?" Jacob asked.

"Naw," the trustee said. "We don't care if you run."

Off the corridor a small delicatessen served up hot food. Bright chrome fixtures stood behind the serving line. Inmates in jeans, sweat shirts and boots walked along the line picking food items from glass cases.

The trustee led Jacob into a hallway, with the same bright yellow ceramic tiles on the walls and ceiling. They came to an opening in the wall.

"There," the trustee said.

9

Jacob entered to find Kelly in an upper bunk. The room was six by eight feet, barely big enough for a steel double bunk, a toilet, and a bookshelf bolted to the wall. There was no door and no window. Jacob leaned up against the bunk and put his head down. The cell was completely barren except for a large old movie poster on the wall opposite the double bunk.

"I need a shower," Jacob said.

"They only give showers between ten a.m. and three p.m."

"Damnit."

"What's it like out there?" Kelly asked.

"Hot."

Jacob turned around, put his back against the bunk's steel frame and looked at the poster. It was a scene from an old Laurel and Hardy flick that showed Ollie dripping wet while Stan Laurel looked on helplessly with a fireman's hose. The poster caption read, 'Here's another fine mess you've gotten me into!'

"Well," Kelly said, leaning over on his bunk and lifting his head up looking at the poster, "Here's another fine mess you've gotten me into."

Jacob saw Kelley's strange little smile.

"I didn't mean it," Jacob spoke in a high-pitched squeaky voice, mimicking Stan Laurel, "It wasn't my fault. I didn't do it!"

Kelly's smile deepened. He came down close.

"It's all your fault," he scolded Jacob. "All of it!"

He stretched out on the upper bunk. "I on the other hand," he lay over on the steel springs, "am as pure as the wind-driven snow."

Jacob slumped into the lower bunk. "Wind-driven snow," Jacob said. "You must be it around here."

He put his hands behind his head thinking of the grounds he had seen today, of the dead animals, of the heavy ash and soot in the air.

"How hot is it?" Kelly asked.

"Hot, but doable," Jacob replied, repeating a code word that he and Kelly had agreed on. *'Doable.'*

"And how could you tell that?" Kelly asked.

"By the age and condition of the deer. This place is dangerous," Jacob confided, "but not nearly as deadly as they want us to believe."

Kelly said nothing for a long while. Jacob knew he wasn't sleeping, but like him planning their next move.

"Well then," Kelly finally said, "It's doable."

Time flowed by into evening. Jacob tried to sleep, but it was no use. Above him Kelly hadn't moved in three hours. Outside their cell the lights dimmed. Inmates moved to their cells. Jacob wondered why none of them had come into the cell. Nor, it seemed, did they talk to one another.

"Nobody talks here," Jacob said softly. Kelly didn't answer.

"I said nobody talks here," Jacob said rising his voice.

"I heard you the first time," Kelly said.

Jacob raised his head as a shadow darkened the entryway to their cell. The lights dimmed again and Jacob made out the outline of the Warden.

"Getting settled in?" Veeby asked.

"Cozy," Kelly answered.

"More trouble?" Jacob asked.

"You have a complaint?" Veeby asked politely.

"The lawn mowers are radioactive."

The Warden squeezed into the tight space between the wall and their bunk and stood in front of the poster.

"The whole dump is radioactive."

Kelly sat up and dangled his feet off the edge of the bunk.

"You're close to that wall," Kelly said.

Veeby squeezed up to the tiled surface with his hands pressed palms out pressing against the walls.

"The walls have lead in them," the Warden said.

"I'm a research scientist. I've studied radioactive waste disposal for over twenty years. I respect it, but I'm not afraid of it."

"You've got guts," Kelly said.

Veeby seemed amused. He leaned against the wall and Jacob could make out his looking from him to Kelly in the dim light

"I'm a chemist by training," Veeby said. "I blew the whistle on the toxic gasoline." He put his hands in his pockets, leaning back relaxed. "I got my choice of jobs. This was the highest paid."

"It should be," Kelly said.

Jacob sat up to get a better look at the Warden. Jacob had the feeling that the warden expected something from them. Jacob cocked his head to one side staring at Veeby.

"Lead in the walls?"

"But we reverse it," Veeby said. "The further inside you go the more protection from the earth and the lead. Inside is safe. Outside is poison. People that misbehave go out closer to the debris. We lose quite a few people that way."

"How can you get away with that?" Jacob asked.

"I'm the fair-haired boy," Veeby answered. "I can do no wrong. In this radioactive wasteland I am king. If I make a policy that's the policy. Nobody questions me. Not even the Governor."

He paused again. He seemed to wait for some response. Jacob sensed that the Warden expected him to say something, but Jacob couldn't figure out what. For another long minute the three men stared at one another.

"You're the king here?" Jacob asked breaking the silence.

"Absolutely," Veeby said. "It's a tough world. Everybody needs to find a niche, a hole. A place where he sets the rules. Nobody hassles me here. My wife is happy. My kids go to private schools. No complaints."

The cell now was nearly completely dark. Figures moved around in the hallway with pale blue flashlights. The light had a strange blue-green quality. Veeby shifted his position along the wall.

"Those lights for example," he said. "Bioluminescence," he said. "The same process that gives a firefly its light."

"Why not use batteries?" Jacob asked.

"Money and security," Veeby answered. "With batteries they could smuggle in a radiation meter and…" Jacob nodded.

He knew at that instant that Veeby was a competent man. Yes, with a way to measure radioactivity an inmate could weave his way passed the really dangerous debris surrounding the installation and simply walk out.

There would be no simple practical way to keep small handheld instruments away from the inmates.

"We have instruments in all the cells that detect the presence of direct or alternating electricity," Veeby said. "Even tiny voltages give out electromagnetic signatures. We take batteries away from inmates all the time."

"Very efficient," Jacob said.

"Well," Veeby said, "I wanted to repeat my message. I do this with all new arrivals. It's really not such a bad place unless you get stupid. Do what you are told and you can serve your time here?"

Again he seemed to hesitate as if waiting for something. Jacob saw him slowly moving toward the entrance when Kelly finally chose to speak. "Yes," he said in a low voice, "That would be doable."

Veeby stopped in the entrance. He stood with his back to them for a long moment.

"Huh," he said finally. He seemed to be unsure of himself. "Then it's you two?"

"Yeah," Kelly replied softly. "It's us."

Veeby came back walking deeper into the cell. He watched outside and when foot traffic cleared he reached into his pants pocket.

"Your man came through," he said pulling out some gold coins. He gave Jacob and Kelly ten gold pieces each and put the rest in his own pocket.

"We get more," Kelly said.

"Uh no," Veeby answered. "I had expenses."

He patted the walls. "Around here I am king," he repeated. "So, when I tell a crew to tear the lead out of a wall that's what they do. Got it?"

"Yeah," Kelly said softly. "When?" Kelly asked.

"Right now," Veeby said. "Why wait?" He patted the tile wall. "This hole will still be radioactive tomorrow."

Jacob knew they were entrusting their lives to this man that already had their money and could take the rest. He could drop them into a radioactive ditch and let them die.

There was something about Veeby, however, the way he had taken his time with them and talked with them. Jacob felt that he could trust him. Veeby went out first, checked the hallway, and gestured them to follow.

They went deeper into the dugout. Hugging the walls they walked deep within the hill. They walked for ten minutes passed open cells. Jacob heard the sounds of inmates moving around. The sounds were low murmuring of men fighting off constant panic.

At night the dugout was like a tomb. There were no radios. No television sets. Only strange blue-green lights illuminating tile walls and sounds of breathing filtered out to dimly light connecting passageways.

They walked silently through tunnels. Veeby stopped and shoved open a heavy metal door. Warm air breezed inside. Now, in the distance Jacob heard the puff-puff of the steam locomotive that seemed never to stop bringing more debris into the compound.

Once outside Veeby held a battery-powered flashlight. They walked around mounds of debris. Lights from New Boston cast shadows over a heaps of debris and twisted metal.

"Touch nothing," Veeby said, "Keep moving. Walk where I walk."

They made their way around grassy hills and walked along narrow pathways. They moved quickly around ponds that glistened in pale moonlight, and smelled like sewage. Jacob heard the low calling of some animals out in the wastelands.

Veeby walked confidently, although Jacob couldn't see much outside the white cone of light. Splinters of light from the surrounding city cast hard shadows on mounds of scrap metal and crushed automobiles.

Coming out around another huge heap of mangled metal they stopped by the railroad track. Steam and smoke surged up over a dark pile of debris backlit by city lights.

As the steam locomotive chugged into view Veeby put his hand up to Jacob's chest. Kelly instantly stopped behind them.

"Steam?" Kelly asked. "I never saw one before."

"We brought them from South Africa," Veeby said. In the dim light Jacob saw the warden smile. He actually seemed proud that America was reduced to buying steam locomotives from South Africa.

"The train is automated," Veeby said in a hushed voice. "There's no human crew or driver. It's hot."

"Who puts coal in the boiler?"

"Robots," Veeby said.

"Robots? But ahh."

"Nobody checks it," Veeby whispered, "because anyone hiding out on it would be dead in about two days. Yeah," he looked up to see the headlights of the locomotive bearing down on them out of the evening mist.

"Here it comes."

The old puffer came along a straight line of track gaining speed. The sounds of hissing steam and a clanking old locomotive came toward them.

"Pay the price and ride the train," Veeby said.

"Ride the train?" Jacob asked.

As Veeby stepped away from the tracks the locomotive came abreast of them and kept going. The locomotive chugged passed them, enfolding them in a bellowing shroud of steam and rumbled it's way up the rail line about two hundred yards. The train, Jacob estimated, pulled a couple hundred boxcars.

Jacob looked to see Kelly standing by the track. His hands were on his hips. His head was forward as if he were intently watching the train.

"South Africa is a haven for steam lovers like myself," Veeby said over the rumbling of boxcars on the rails.

"You should go there sometime. I did on my last vacation."

He frowned, "But of course things are not so good there either."

"We're safe standing here," Veeby said. "But remember the cars are radioactive. So jump up. Find the box and get inside. Close the lid down. There's a lever inside the box. Got it?"

"A box?" Jacob asked. A sense of dread pulled at his heart.

"Lead boxes inside the boxcar. Air and water inside the boxes," Veeby said. "You'll be safe, but don't use all the air."

Veeby took a small device out of his shirt pocket, aimed it down the track and immediately the locomotive vented steam into the night air as the old engine slowed down.

Dozens of boxcars rattled by and finally, a quarter mile down the track the engine wheezed to a stop.

"Yeah, that's it. Car Bravo Foxtrot Five," Veeby pointed. Huge black letters painted on gray metal door read BF5. A sliding door on the boxcar opened automatically.

"Remember," he said, "It's hot. Nobody will check inside. Don't look at the view. Go now. Go!"

Kelly dashed for the boxcar, grabbed a handhold and swung up into the dark interior. Jacob gulped and felt Veeby give him a little push.

"Go!" Veeby said. "I can't hold the train!"

Jacob stood frozen with fear. Standing on the tracks surrounded by radioactive waste and staring into the black lifeless interior of the boxcar, he couldn't move. Sweating in the hot Indian summer night Jacob's mouth was parched. His feet felt like lead. He heard something slam shut from inside the dark boxcar.

"Kelly?" he called.

"Go!" Veeby whispered harshly.

Before Jacob could move Veeby ran back toward the installation. Jacob turned, but already the Warden was out of sight. Jacob only heard fleeting sounds of Veeby running along with the sounds of the locomotive engaging.

A hiss of steam vented from a valve as the locomotive chugged to build a head of steam. Boxcars rattled as the undercarriages took up slack.

"Kelly?" Jacob raised his voice.

In the distance Boston glowed in the night. City lights reflected off of huge heaps of radioactive scrap metal. The boxcar started to move and Jacob ran desperate not to be left behind. He grabbed a handrail.

Scrambling up into the car Jacob felt its oppressive warmth as darkness closed around him. Dim blue lights shone up from inside an open lead box. Jacob leaned down, putting his hands on the sides of the crude massive lead coffin and looked down as two 'firefly' flashlights blinked on and off in the bottom of the lead box.

Jacob saw three oxygen bottles lying along the sides. The interior was about his size and, like a casket, was lined with a cheap white cloth. Once inside Jacob fumbled around for a control level to close the lead top.

Rolling over he saw a dirty pillow, a small canister labeled H2O, a second water bottle and the three steel air bottles labeled Oxygen.

Jacob pressed a button on the side and two hydraulic pistons wheezed the heavy lid down on top of him to seal him His hands pressed up against the heavy top, but he couldn't lift it.

He guessed the lid itself weighed a thousand pounds. He pressed the button again, to test the hydraulic pistons, but the heavy lid did not move. Terrified, Jacob knew he was entombed. In the box he fidgeted with the flashlights. Setting them on his stomach, he steeled his nerves.

"Using too much air," he said. "Damn. Don't talk."

Putting his ear over as close as he could to the wall of the lead box Jacob listened, but he could hear nothing. The box swayed side to side and he knew the train was moving. Down at his feet the air bottles went pizz-pizz releasing precious oxygen into the cramped space.

Jacob lay paralyzed in fear. How long would the air last? He couldn't lift the lid himself and wouldn't know when to lift it. His life depended on others. His life depended completely on Veeby: If he had chosen to do so Jacob knew he might die inside the coffin and nobody might find him for years. Slowly he got his breathing under control. He switched off the light, then immediately turned it back on; because the darkness was more extreme than anything he had ever experienced.

Rubbing his fingers along the underside of the cover he felt rough edges. This was what eternity was like for those who left Stonegate dead, except he was still living. Huh, yeah, he said to himself, that's what they say. Only two ways out of Stonegate: By air or in a lead box.

Jacob had no sense of time. He may have slept. He didn't know. His bladder filled and emptied many times. His butt squished when he wiggled around trying to get comfortable. His back was killing him, but there was nothing he could do about any of it.

His eyes and mind gave out. Jacob didn't know if his eyes were open or closed. The darkness was complete. Nor could he tell anymore if the train moved. His brain couldn't tell the difference between the train moving and the movement of his own body in the tome.

It might have been hours. It might have been minutes, but then mercifully there was glorious wonderful light as the lid swung open. A person in a white radiation protection suit dropped an object down on top of him. The object was white and big. Slowly, Jacob recognized it as another protection suit.

"Move!" the man shouted.

Two men reached down pulling Jacob bodily out of the box and shoved his legs into the bottom half of the suit. He scrambled into the rest of it. Jacob was half dragged and half carried out of the boxcar.

He glimpsed the train pulling away from a huge tin shed. A sliding metal door closed. A light blinked on. Jacob's eyes adjusted to the interior: The walls were of lead. Two men pointed to a sink in the back and over to a rack of suits.

"Clean up!" one commanded.

As they walked away Jacob slowly got out of the protection suit and saw a shower stall across from a toilet. Clothes swayed on the rack. Someone was behind the rack and shoving on it as if rifling through the clothes.

"Ebbtide?" Kelly asked. He came out from behind the rack. "I thought we left you behind."

"I'm here," Jacob said.

Jacob stripped, dropped his prison garb on the floor, went into the shower and washed. He stumbled to the sink on wobbly legs. He washed his face three times and shaved with gear laid out neatly on a table.

"I'm here," he repeated as he came out to inspect the suits.

Kelly stood to one side watching him. He was already fully dressed. He wore a very smart silk shirt with a red tie and brown shoes. Kelly studied his image in the mirror. He looked like he stepped out of a fashion magazine.

"Quite a selection huh?"

Kelly pointed to the rack. Jacob eyed a dark blue suit. A white tie with dark blue strips. He selected a light blue shirt to complete the selection.

He selected a set of black loafers like lawyers wear. When he was dressed he felt like a new man. Thirty minutes later the two men stood side-by-side fully dressed and feeling dapper.

Jacob looked around the barren room filled with what were probably stolen clothes. This was a gang hideout. Jacob had to admit it was safe. The whole warehouse was likely radioactive. Maybe it had been used to store waste materials before the debris was loaded onto the train. Jacob wondered idly if maybe he should invest in lead mining.

"Okay gents," one of the men said. "Let's go."

Outside it was daylight. Once in the sunlight Jacob felt disoriented. He was exhausted from the train ride, but the change into new clothes and the shower refreshed body and soul.

"Where to?" Kelly asked.

"The harbor."

Jacob was dizzy by the time they reached the water, because he needed food. It bothered him that he couldn't figure out if it was sunrise or sunset. Nor could he tell what part of the harbor he was in. There was glorious sunlight, but the sun was low in the sky, a deep orange.

Long shadows fell across cargo ships and gantries lining the waterfront. Dozens of old abandoned warehouses sat side by side as if patiently waiting for good times to return.

All along the shoreline rusting cranes and derricks sat in semi-darkness squatting like huge metal frogs. Here and there a building had lights on.

Jacob saw shadows fall over the harbor. In his disorientation a few of building windows seemed to reflect shadows rather than light. The glass got darker as the setting sun swept downward on the horizon.

Jacob suddenly knew that he must have spent a night in the train, lying in the lead box, waiting, waiting, for the men to rescue him. He realized he had spent a night and gone less than three miles from Stonegate to the harbor.

A gleaming white yacht sat at anchor out in the harbor. Its sleek lines and trim chrome fixtures spoke to Jacob of old money and the good fortune and of those lucky enough to lead that kind of life. He felt exhausted and weak, but tightened his tie and straightened his coat.

"Down the ladder," a man said. Kelly went down first and stepped into a small motor launch.

"Go on," the man said as Jacob hesitated by the dock ladder. "They are waiting for you."

Jacob knew the yacht probably belonged to a Kamsi royal family member. Harbors around the world were filled with them. Jacob made it down the ladder and sat next to Kelly as the launch motored them out to the yacht.

A servant opened a hatch. Jacob and Kelly entered a gracious dining room. As the door closed behind them Jacob heard their escorts walking down the gangplank. His eyes adjusted to the interior lighting. He saw a feast set out.

Silver plates of figs, dates, fruits, iced cakes, lamb and beverages piled the center a long rosewood table. Jacob walked in to the table, stopped, looked at the food and wiped drool from his mouth. He had never seen food look so good.

The dining area was cool, but not cold. A gentle air vented the dining lounge. Servants entered by a far door as did Sheik Kasan.

"Welcome Doctor Ebbtide," the Sheik said smiling and extended his hand in welcome. "I had a feeling we would meet again. Thank you for contacting me with your plan."

Jacob warmly accepted his handshake. He introduced Kelly. The two men smiled at one another. Kasan nodded and servants held chairs for them. The Sheik gestured for Jacob and Kelly to sit. Hot food was served.

They ate in complete silence as was the Kamsi custom. Jacob ate with his right hand only. He ate until he thought his stomach would explode. He burped long and loud. Even the servants smiled approvingly.

Afterward bowls of water were brought to each person for washing hands. Ten minutes later found them in a luxurious sitting lounge. Deep leather seats reclined automatically as the three men settled down. Kelly laughed out loud, startled, as he was taken to a flat position by the electric seat. Kasan got up touched the controls as Kelly rested his hands on deep padded armrests.

"You got some digs here," Kelly said.

He made small talk with the Sheik for a moment and then Kasan moved away ending the conversation. The Sheik stood by a window looking out at the harbor.

"We have an interest in a contract with Dr. Ebbtide," the Sheik said.

Kelly sat for a moment, nodded, mumbled something and started to get up, but fumbled with the seat controls.

"We will help you," the Sheik said. "We appreciate what you did."

"Wasn't much," Kelly replied, "I used to be a cop. I know the system. The way things are today anything is for sale even freedom."

"Always?" the Sheik asked. "Is freedom always for sale?" Kelly's eyes met Jacob's.

"No not always."

"And you?" Kasan asked.

"Not for me either," Kelly said. "Freedom is the wrong word I guess."

The Sheik stood by a large dark-tinted window with his back to Jacob and Kelly. "You meant that escape could be purchased. Is that way you meant to say?"

Kelly took a deep breath. He moved the seat controls and sat up straight. "I will need help to stay clear," Kelly said. Jacob detected a tone of pleading. Kasan nodded and turned to Kelly.

"We will help you," he repeated.

Jacob suddenly felt sorry for Kelly, because even with Kamsi money he would always be running. The legal system in the United Remnant was mortally wounded, but not dead.

"I have some business to take care of," Kasan said. "I will assign a man to drive you wherever you want to go." Kelly stood up and shook hands with the Sheik.

"Thank you Sir."

Kasan looked from Kelly to Jacob. He seemed to understand that friends need time to say goodbye.

"I will give you a few moments," he said.

When the Sheik left Jacob and Kelly looked at one another. Kelly said nothing, but looked out at the harbor and the fading lights of the city.

"Tell me something," Jacob asked, "Why was your wife divorcing you?"

"You know," Kelly said.

He nodded. Yes he understood. Kelly would not be welcome in the Kamsi Kingdom. They have a strict code.

Men marry women. That's it. That's the system. There are no exceptions. Men who prefer other men are not tolerated.

"And the con about the gasoline coupons?"

"She was getting back at me. She felt betrayed."

Jacob faced Kelly. "You wouldn't be happy there," Jacob said.

"And *you* will?"

"I can't spend my life running. I am a scientist," Jacob said. "I need to move freely in whatever society I am in. I can't spend my life running."

"I can't either," Kelly replied softly. "The trouble is I don't know reliable people. I know criminals. I know police, but nobody I can trust."

"There is a man," Jacob said. "He's a brother of a good friend of mine. It's Zada. Doctor George Zada. He's a brother of my best friend."

Kelly nodded waiting. Jacob had the impression that Kelly would welcome contact with anybody that was trustworthy.

"He's a scientist like me," Jacob said. "I think he's still in the country. In fact I'm sure of it. You can check the national phone directory. I'm sure the phone number is right. You can get his street address either off the Internet, one of those personal databases or from a reverse directory service. Tell him that you're a friend of mine. He'll help you. I'm sure." Jacob leaned over and whispered the phone number.

"Have it?"

"Zada?" Kelly asked.

"He's a scientist," Jacob repeated.

"Can he help? What does this guy do?"

"The last I heard, but this was about two years ago, he was working with insects. He had trouble. He's a loner and he's probably crazy, but he's one of the most honest men I ever met. He hates authority."

"Was he your backup plan?"

"He was my main plan, but I got nabbed at the airport. I never had a chance to contact him. My goal was to avoid getting caught. Maybe he could have helped me with that, but maybe not. Zada is a question mark. I know the Sheik. For me this is a better choice."

Kelly put his head down closer. "So?" he asked.

"What am I supposed to do?"

"Look, maybe you don't make it," Jacob said. "I mean, you know, suppose the worst happens and they nail you. I'm saying that this guy doesn't get along with authority that's all."

"Oh, he takes after you. Is that it?"

Jacob frowned, but lifted his hand in a farewell handshake to Kelly. "Good luck and thanks."

"All the best," Kelly said. As they shook hands Jacob added, "Be careful with Gladys. She's dangerous. If you get caught try to get back in her good graces."

Sheik Kasan came back into the lounge. He stopped by the table to pour a drink while a servant handed Kelly a small leather pouch. Kelly's fingers trickled gold through them.

"Very generous," Kelly said.

Kelly gave Jacob a weak salute and followed the servant out of the lounge. The door closed behind them and Jacob heard Kelly's footsteps on the deck.

"What about your feet?" Jacob called after him, but Kelly didn't hear. The Sheik handed Jacob a drink that tasted alcoholic like date wine, but stronger.

"I didn't think Kamsi drank alcohol."

"When in Rome do as the Romans," Kasan said. "Isn't that what they say?"

Jacob and the Sheik toasted one another. When the Sheik put his glass down on the table Jacob did likewise.

"We need to talk," the Sheik said.

The Sheik turned on a light. Outside the world was dark, but inside the lounge only a puddle of yellow light kept the blackness outside from enveloping them.

The date wine was smooth and deceptively strong. Jacob felt his tired legs relax. He eased into the seat and listened to the Sheik.

"You need to understand, Doctor that things have changed in the Kingdom since you were last there."

"Changed?"

"It's more hostile now. Worse than even two years ago."

Kasan spoke at length about the increased hatred of foreigners, of the Rules of Conduct, and how some of the Kamsi leadership was trying to bring a frantically religious population into the new century. "It is quite hostile," he repeated. "I want you to know."

Jacob put his head over thinking. "Yes," he said at last, "I know things have changed, but if rules are obeyed is there still a problem?"

"It's always a problem Jacob," the Sheik explained. "You have a right to know what you're getting into. That's all. I want you to know."

When Jacob said nothing the Sheik poured another drink for Jacob and one for himself. For a long moment neither man spoke. Both held their glasses while staring into space. Each in their own thoughts.

Jacob sensed his host's concern and felt that a bond of friendship had developed between them. Jacob was content to stay where he was. After a long pause the Sheik leaned forward while setting his glass back on the table.

"You have a daughter. Her name is Talya?"

"I think my wife knows something," Jacob said. "Talya is missing. She's been missing for over a year."

"We believe she is in the Remnant," Kasan said. He leaned closer. "But now Doctor you must swear that you know nothing of her whereabouts or her activities."

"Whereabouts?" Jacob asked. "She's been missing. No ransom was made. Does Leviman know where she is?"

"We don't think so."

A thought about Ben Leviman occurred to Jacob. "Is he still President?" Jacob asked.

"He made himself President for life last year."

Jacob shook his head. Israel was becoming another third world country with presidents that elected themselves. After a moment he looked up as Kasan spoke.

"We think she is in the Remnant," Kasan said again. When he said nothing else Jacob sat his glass down and looked at the Sheik.

"Tell me," Jacob said. "I want to know."

"I will explain in time," Kasan said. "You need to know our operations..."

"What do your operations have to do with my daughter?"

"She is interfering!"

"With your operations?"

"Yes."

"That's impossible!" Jacob shouted. "She's only nineteen. She's probably captive somewhere!"

"Maybe she is being used," Kasan said. "Used by the Americans."

Jacob knew Talya was probably safe, because no demand had been made for a ransom. Maybe she was in the Remnant with Herman and Emil. Maybe the Americans had the missing Israeli scientists to punish the Kamsi, but what would Talya have to do with any of that?

When the Sheik sat quietly staring at him Jacob turned the situation over in his mind. One hostile thought reoccurred. Both Herman and Emil were his top scientists and when they vanished long ago Jacob was sure his operation in Tel Aviv was thrown into disarray.

Was Leviman or people in the Remnant still out to get him? Jacob couldn't get his mind around the politics. It seemed that everybody was gunning for him even to the extent to claiming that his teenage daughter was causing harm.

"Tell me," Jacob said.

"I need to ask," Kasan said. "It is a formality."

"I doubt that my daughter could do you harm. Even if she wanted to."

"It is a formality," the Sheik said again. "You know," he said, "People believe that societies are governed from the top down. They think that there is leadership, but in the modern world that's not the way it is."

"People at the top react to what's going on. That's all they can do," Jacob replied.

"It is the bottom tier of society that drives the process," the Sheik said, "Especially in a desperate world like ours. It is, as you might say, the huddled masses that call the shots. The leaders are fighting for their own survival."

"Yes. There's truth in that observation," Jacob said.

"In Kamsi society," the Sheik added, "There are those like myself that believe moderation, tolerance, is essential if we are to compete in the modern world."

Jacob nodded. He understood that Kamsi society was ruled by Sheiks, local men with power and that there was no 'Rule of Law' other than local interpretations of Scripture.

Even local chieftains had to stay in step with their rank and file. Jacob he felt emotionally secure so long as he could count on men like Kasan.

"I understand, but I can't stay here in the United Remnant," Jacob said, "and I'm not welcome in Israel."

"I know," the Sheik said. "That is regrettable." They sat for a while sipping date wine. When Kasan spoke he seemed bothered and confused.

"If you can tell me," he asked, "Did you speak with President Leviman?"

"No," Jacob said sourly. "He never called. I let people know the best I could that I was in prison. Leviman could have found me. If he wanted to find me."

"I had the impression," the Sheik said, "that he just wanted you to come back to Israel. I guess I was wrong. He seemed very anxious about you. He seemed completely sincere. It troubles me."

"Why?" Jacob asked.

"Because I like to think I am a better judge of people. You can still go back."

"So he can kill me?"

"I doubt that it was ever his intention. He sent you to us to see if you would return. At least that's what I understood."

"He told you that?"

155

"Yes. If you wish we can still make arrangements."

"Leviman is a dangerous man," Jacob said. "He's a criminal. And, he's a liar. I can't go back to Israel."

"Then you are most welcome to work with us," Kasan said. "I hope so."

"You'll be well paid Doctor. An account will be opened for you in a Swiss bank, but remember my friend when in Rome do as the Romans."

"I accept," Jacob said, "and I will."

Kasan moved closer and took Jacob's arm affectionately helping him stand up. A silent agreement passed between them.

The Sheik seemed satisfied. He made a gesture for Jacob to follow him. They went down a flight of steps to another deck.

A room was setup with a central display table. A lighted map of the Middle East filled the entire area of the display. The table was a geographic display that showed the whole region from the Mediterranean Sea on the north and west to the Kamsi Kingdom and the Arabian Sea on the south and Iran on the east.

Then Jacob noticed he had overlooked a large part of Africa that extended in a broad swath south of Egypt down through the Sudan and west to Chad. Jacob stood for a long time studying the new geography, letting the new geopolitical reality settle into his thinking.

"I had no idea that the Kingdom had extended so far."

"Millions of square kilometers," the Sheik said sweeping his hands over the map. Jacob sensed the Sheik's pride at the extent of the Kingdom.

"Millions of square kilometers," the Sheik said again, "but frankly, Doctor, the land is almost all desert, bone dry, desolate."

"Solbean is the answer," Jacob said. His gaze took in the vast map and his imagination took flight at the tremendous possibilities of matching a marvelous genetically engineered plant like Solbean with the vast sunlit stretches of desert sand. Jacob saw Kasan watching at him.

"We've planted almost a half a million square kilometers," Kasan said. "We know the possibilities."

"The old Solbean?" Jacob asked.

"It grows," the Sheik answered. He frowned.

"And it dies?" Jacob asked.

"Growth is poor and spotty. Yield is poor."

"The Israelis are selling you seed?" Jacob asked.

"We buy from the Israelis," Kasan said. "But another source is the Americans. We paid millions of gold dinars to help settle claims against our poisoning the oil. We were wrong to do that, but we felt we should get some goodwill in return.

"We were wrong about that too. We thought the Americans would sell us good seed."

"The Americans said so? They lied?"

"No. It was Leviman. All the seed from America is inventoried in Israel. He said he had brokered a deal for us through the Americans."

Kasan reached under the tabletop display and flipped a switch. A bright red line illuminated in the display.

"Seawater is brought in from the Red Sea," he said, "and mixed with a small amount of fresh water to irrigate the crops. Solbean is truly marvelous. Did you know it could grow in nearly all seawater?"

"Yes," Jacob said feeling a sense of pride that his work and that of other talented people made possible such strides in agriculture possible.

The Sheik gestured to the display. "The desert doesn't give up easily. We need you to advise our engineers on how to stop the Sahara from encroaching on our crops and of course there's always the poor quality of the seed and the Americans."

"American satellites?"

"They are back in orbit," Kasan said.

"Same frequency?"

"The same. We play cat and mouse with them. They want strict control," he said. "I guess they don't trust us. Even though we purchased the seed through the Israelis at a great price. They have recoded their satellites since you damaged them two years ago, but I think they are really afraid of you."

"I doubt it," Jacob said. "I was in jail in New Boston. They could have found me. The government has no interest in me."

"Yes they do," Kasan answered, "According to the information we have your lady, Judge Hines, kept you to herself. We think she never put your name in the national directory. I guess she had her reasons."

The Sheik hesitated. "We don't think the Remnant government ever actually knew you were incarcerated. I also think that the Israelis didn't know," Kasan said, "Because I don't think your wife told them either."

"Anything is possible with my ex-wife."

Jacob thought for a moment. He tried to understand Gladys. She might have informed the Remnant government. Or not. Jacob found it difficult to believe that nobody came looking for him. The Hines Prison House was two miles from the spot he was arrested.

Gladys had authority, but that authority was limited. If Kasan or other had notified the Remnant government that he was coming into the country the government could have investigated. They could have found out that he was in Boston as an inmate at the Hines Prison House. It would not have taken much investigation to find him.

But the fact remained that nobody from the government had attempted to rescue him from Gladys. For her it was clear sailing and for Jacob it was lawn mowers with brakes.

"Could you not make contact with them?" Kasan asked.

"I had limited use of a telephone," Jacob said, "but the numbers I could call were limited."

"Was there no way to get word out?" Kasan asked.

"I knew a man named Handy years ago," Jacob said.

"He was a security consultant. He was still in New Boston when I was jailed. He remembered me, but he seemed to think that I was already working for you."

"For us?" Kasan asked. "Why would he think that?"

"I think Gladys has her tricks," Jacob said. "She has people that can fake the phone numbers."

"Yes, the Chinese are clever that way. Dialing a phone number is now like gambling. Maybe you connect to the person you are calling, but maybe you don't."

Jacob nodded. He noticed a quick smile cross the Sheik's face and then almost instantly vanished as he caught Jacob looking at him.

"This consultant saw a return phone number from the Kingdom?" Kasan asked.

"I don't know," Jacob said "I know he didn't believe me. I spoke with him once and the staff cut that phone number."

"Cut the number?"

"My wife had control," Jacob said. "She could delete certain phone numbers."

"Delete the number?"

"Pull it out of the system."

"And the consultant?"

"I spoke with him a couple times, but couldn't convince him that I was in New Boston."

"Couldn't convince him?"

"Gladys, my ex-wife, didn't have me in the system. Not officially. Handy goes by the book. He made inquiries, but…"

"Your wife kept you out of sight and off the books…"

"She's a judge," Jacob explained. "She knows the law and people take her world for things. She lied and kept me out of the system. I never left the premises."

"No visitors?" the Sheik asked.

"No. Not even from the government."

"You couldn't convince them?"

"My phone calls were all routed through a Middle Eastern network."

"I see," Kasan said softly. "Yes, the phone network is through China. It is their tool. Money is king."

To others Jacob was calling from someplace in the Middle East no matter what he said. It hurt him to think that he couldn't convince anybody that he was in prison in the Remnant. His word was good for nothing.

But of course Handy would have also talked with Gladys. She would have told them another story all together. Jacob? Here? That's ridiculous. Well, he never was a very honest man. That's why I divorced him.

"It was her word against mine," Jacob said.

"She *is* a Judge," Kasan replied.

Jacob understood that he was being diplomatic: Everybody knew about the O'Hannon gang and the Israeli Intelligence Services. People understood that in many types of work lying is part of the business and people in those businesses sometimes lie as a matter of easy habit.

"With my past I understand," Jacob said.

"You wear many hats," the Sheik said softly.

"It costs me credibility with people," Jacob said. He felt sorry for himself. His word was assumed to be a lie. He took a deep breath and let his mind concentrate on the magnificent table display.

Kasan stood on the other side of the table studying Jacob. "It would not have been in our interest to intervene," Kasan said. He sounded apologetic.

"At least not until you came to us with a request for our help."

"I understand," Jacob said. "You could have turned me over to the government, but..."

"You still have that choice."

"No," Jacob said. The government in Washington was in constant communication with Leviman. If anybody wanted to find him Jacob knew they could. They didn't look.

As for Kasan, well, Jacob thought, everybody is playing his or her own game. Everybody has an agenda. He couldn't blame the Sheik for faithfully serving his masters. Jacob shrugged off the fact that Gladys had hidden him from the government. His attention went again to the display.

He saw outlines of what used to be other countries, including Algeria and Egypt. The Kamsi Kingdom had vast real estate. And, thanks to Solbean the Kamsi had a plant that would grow almost anywhere producing huge amount of biomass for conversion into liquid fuels.

"I can help you," Jacob said confidently,
"Especially with the American satellites."

"And the seed?"

"Yes," he answered confidently. "I can produce superior seed for your operations."

With these words the Sheik flipped another display switch and another network of lines illuminated in the display.
"We are building a railroad," the Sheik said pointing to an extensive layout of track.

"You see," he explained, "When we built a pipeline that feeds seawater to the Solbean we built a railroad at the same time. But of course…"

Jacob looked at the radiating set of lines representing railroad track. He was puzzled by the fact that much of the lines didn't seem to connect. There was nothing between the tracks, but sand.

10

Somehow a locomotive would have to move from one set of tracks to another. He raised his hands. The Sheik pointed to the wall behind him.

A wide screen television lit up filled with images of a strange railroad apparatus. The video clip showed a train-like machine consisting of hundreds of combination steel and rubber balloon wheels traveling over steel rails or over sand with equal ease.

A long freight train pulled dozens of tanker cars. The locomotive had oversized rubber wheels. The railroad cars ran on a combination of rubber and steel wheels.

"What is that?" Jacob asked.

"We call them Centas," the Sheik said. "You know after a famous desert insect the centipede. A creature with hundreds of legs."

"It looks like a centipede," Jacob said, "The way it travels over the desert." The machine undulated over the dunes as it wheeled along on hundreds of over-sized tires.

"Each railroad car is self-propelled," Kasan said. "The Centas can operate on steel rails or directly over sand. It can travel up and down the dunes with ease."

Jacob studied the images of the Centas and how, when it came to the end of the track, it kept going. It didn't even slow down.

"No driver?" Jacob asked. The video showed the engineer's cab of a locomotive. The cab was empty. Nobody was driving the lead car? Jacob looked around to see that Kasan stood on the opposite side of the display table. He reached under and turned off the light. The map of the Kingdom fell into darkness. Suddenly, with that question, something had changed in the Sheik's demeanor. Jacob suddenly felt terribly alone. He face the Sheik across the darkened map

"No driver?" Jacob asked again.

"No," the Sheik said. "The trains are automated. They travel on tracks and over desert sand. They are controlled by Ground Positioning Systems and by signals from satellites."

When he stopped talking he stared at Jacob as if waiting for an answer, but Jacob was simply confused. The Sheik turned the set on and moving images again filled the screen.

Jacob could understand automated railroads. So what if the locomotives didn't have drivers? This is a modern age. Many machines are controlled from space.

Irrigation systems in remote areas are switched on and off after a satellite monitors the moisture in the crop leaves by measuring the color of the radiation reflecting off foliage. The color determines whether or not the plants need water. If they do, the satellite sends a signal for the irrigation system to pump water.

"Okay," Jacob said at last. "The locomotives don't have drivers," he hesitated and asked, "Is there a problem?"

"We don't know," the Sheik said and his tone of voice was now brisk, businesslike. Without waiting for an answer he asked, "Do you see Doctor this is where the problem is with your daughter?"

Jacob faced the wall television set with its images of a driverless train careening off of a steel railroad track and heading off into the desert. Jacob looked back again at the Sheik.

"What are we talking about?" he asked.

"Trains and daughters," Kasan said.

The Sheik walked around the display table to confront Jacob.

"You don't know?" the Sheik asked his voice louder.

"You tell me as a man of honor, Doctor, that you don't know what's going on!"

"I swear," Jacob whispered. "My daughter has been missing with my friends. Since before I was jailed." Jacob looked across the table at the Sheik, his expression bewildered.

"I swear it," Jacob whispered.

"By whom do you swear?"

"By God," Jacob replied earnestly. "I swear by God."

"You are a man with knowledge of satellites. Did your daughter contact you while you were jail?"

"No. I haven't talked to her."

"And your wife?"

Jacob wondered why Gladys didn't seem worried about Talya. Gladys never mentioned their daughter other than to accuse Jacob of kidnapping.

"I don't know what happened to her."

"Somehow," the Sheik said, "What is happening to our satellites bears a strange resemblance to what you did with the American satellites. Am I making myself clear?"

"Your transportation system is attacked?" Jacob stood dumbfounded. "What's this to do with my daughter?"

"She is behind the attacks!"

"She's nineteen!" Jacob shouted. "She doesn't know one end of a satellite from the other!"

"You didn't pass her or Leviman information how you took control of the American satellites? How to sabotage our chemical and transport system?"

"No," Jacob said. His hands hung limply at his sides.

"I swear it," he mumbled. "I never gave information to anyone. I swear it."

"You see our problem don't you?" the Sheik asked.

From Kasan's point of view if Doctor Ebbtide might chose to give information, technical advice, as a way to help himself with the Israelis or the Americans. To gain goodwill.

It seemed logical to Kasan that Jacob would give that information to someone he really trusted and who better than his own family?

When Jacob said nothing the Sheik pointed to the images of the train. "That is the system," he said. "That is our system!"

The Sheik took a small remote device out of his pocket and clicked it at the television. The images changed to one of the Centas pulling into a vast Solbean growing field. Workers in the field harvested the Solbean, carting biomass to processing tank cars.

The foliage was dumped into a hopper on the tank car and then an animated graphic illustrated how the raw Solbean foliage was processed in the tank car into alcohol.

"Oil has to be transported," the Sheik said. "It takes pipelines to transport oil, but oil is in short supply. Solbean is the new energy source. A way had to be found to transport it out of the fields."

"Yes," Jacob said. "I'm impressed. Really I am. Obviously you need a way to get the bio-fuel chemically processed and transported to market. You can't let it stay in the fields and rot. Centas are the answer."

The growing of the plant, the harvesting, and the processing into fuel were all done in the fields. Once processed the liquid 'extract' from the Solbean was piped to another tank car for storage and transport. The unused portion of the Solbean was deposited onto the desert as compost to condition the sand. Over the months and years the sand would become soil. The Solbean would hold the soil in place; anchor it, from the incessant wind.

"Question," Kasan asked. He flipped off the television and walked around the display. "Who do you think recruited her?"

"I don't know," Jacob said, "I'm not sure she was recruited. Maybe Leviman. I don't know."

"What about Leviman? Is he working for the Americans?"

"I don't know," Jacob said, "but I think something is wrong there. I think Talya is a pawn. Are you sure? You think she is involved in the sabotage of the satellites?"

"Almost certainly yes," Kasan said. "She was trained as an engineer at the University of Tel Aviv?"

"She's only nineteen," Jacob answered. "She's smart, but still a student." Honestly," Jacob said, "She doesn't have the expertise to sabotage a satellite."

"Who does?"

Jacob knew that the Sheik meant 'besides him' who could master the cyber code, the programming and master electronic countermeasures to gain control of a satellite?

"Why do you think it's my daughter?"

"Her name!" Kasan replied harshly. "We hear her name used when we intercept electronic messages between the Americans and the Chinese."

"In the open?"

"Yes," the Sheik said. "Her name is used in the open."

Jacob knew both men. Both Herman Burger and Emil Zada were first-rate scientists. Either of them could work their technological magic on a Chinese launched satellite, but the two of them working together would be an unbeatable combination.

Was Talya with them? Jacob wondered if his daughter was with the missing scientists or in communication with them. And, if so, why was her name used 'in the open' when the Americans or the Israelis sabotaged Kamsi operations?

"Somebody is causing trouble with our control signals to our remote trains," Kasan said. "We have Centas scattered over a half a million square kilometers! They are disappearing behind sand dunes or simply driving into the Red Sea! We must bring in underwater divers and salvage equipment! We are facing ruin!"

Jacob felt dizzy at the magnitude of the problems facing the Kamsi. They were not a technological people. In fact, some of the remote tribes were barely out of the Stone Age.

The Kamsi were losing control of a mobile system that must have cost them hundreds of millions of gold dinars to have the Chinese design and build the trains for them.

"I can't believe Talya would have anything to do with sabotaging your system," Jacob said adding, "Even if she could. Are you sure it's my daughter?"

"She is in the thick of it!" the Sheik insisted. "You know the Chinese control most of the global electronic networks?"

"Yes," Jacob said. The Chinese control most electronic communication systems. The Remnant was now a backwater third-world player as are most nations.

The Kamsi had agreements with the Chinese to monitor all telephone traffic and to record certain names and phrases.

The Americans used to do the same thing as part of the War on Terror except that now that the Chinese did all of that work for the Americans as well.

That was how the Kamsi came to suspect that Talya was involved in the sabotage of their satellites: The Chinese were giving them recorded messages between her and others.

"You think the Americans are playing this game on us?"

"With the help of the Chinese, yes," Jacob said.

"They are using my daughter as a way to punish me. It makes trouble for me by using her. They think I'm in the Kingdom."

Jacob rubbed his eyes. "Those dirty bastards!" he swore.

"Sir!" Kasan shouted. "No profanity! You have been warned!"

"It's the telephone system," Jacob said ignoring the Sheik's outburst. "That's what the telephone system was telling them!" Jacob insisted. "Handy told me the same thing. He thought I was calling from inside the Kingdom! Not the Remnant!"

"And the Judge?"

"She denied everything!"

Gladys had him isolated and had probably worked some deal with the Chinese to make sure calls into or out of the Hines Prison House were routed through a Middle Eastern exchange.

"Doctor," the Sheik said, "Once you are in the Kingdom you will know our system. Then, should you turn on us, the damage you could do would be complete disaster!"

"I will not betray you," Jacob said. "My world has closed around me. Leviman has used my daughter as a pawn to make trouble for me. My ex-wife has sabotaged my efforts to work productively. I owe them nothing!"

"I had to be certain," Kasan said.

"Of course," Jacob replied. "I'm not surprised about the Americans, but that he would use my daughter to make trouble for me means I can't go back to the Remnant. I can't go home to Israel either."

A wave of relief swept over Jacob as the Sheik put his hand on his shoulder. "I'm sure that you are sincere," he said. The Sheik's words came as cool water to a man suffering in desert heat.

"I will work hard for you," Jacob said. "The world is desperate for energy. Solbean is the answer. I will help you. But I am telling you my daughter don't have the skills to tamper with satellites.

"If somehow she is involved and if Talya is working for the Americans I will stop it. I swear it. I will stop their sabotage!"

"I believe you," the Sheik said.

Kasan understood the Doctor's situation. The Americans were apparently using his daughter to make trouble for him, because they believed he was already in the Kingdom. By using the girl they sought to create suspicion against the Doctor. Such a tactic could only succeed if the Americans and the Israelis honestly believed that the Doctor was already working for the Kingdom. And, if he had been in the Kingdom for over a year, he was accepted and working as part of the Kamsi team. Kasan saw the American strategy as confirmation of Ebbtide's claim—that nobody came looking for him. They hadn't come looking, because the Americans were already sure they knew where Doctor Ebbtide was—in the Kingdom. The circle was now closed.

Jacob once again he felt keenly aware of his inability to deal with people. Did he really trust Sheik Kasan? Jacob had only his word that Talya or the Americans or the Israelis were committing sabotage against the Kamsi.

It could be that the Kamsi were simply incompetent and casting about to lay blame on others? Yet, he was now painfully aware that he needed the Kamsi and the Sheik's goodwill. Telling Kasan his suspicion that the Kamsi might be engineering incompetents and very possibly their own worst enemy was hardy the sort of thing to smooth his transition into their employ.

Yet, if he agreed to work for money what type of evidence could he provide his employers that they were, in fact, incompetent? How would they react if given positive proof that they were Stone Age people trying to engineer systems far beyond their ability?

Also, Jacob wondered what type of proof would be good enough to show that his daughter had nothing to do with the problems of their agricultural harvesting system.

What kind of proof could the Kamsi accept? Still, even worse, what if it were true? Could he take money from them as well as ask for Talya's life?

"Do we have a deal?" Kasan asked.

"Yes," Jacob said. "I have questions, but for now we have a deal. We will work our way through our problems."

The Sheik poured fresh drinks and they sat until long into the night talking and dozing in the electrically heated deep leather armchairs.

Jacob drank date wine and finally toward dawn he drifted off to sleep. The yacht swayed in heavy Atlantic swells. Jacob roused and made a mental note that the ship was underway. They had motored out of New Boston for a new anchorage in the New Democracy of Maine.

Once there they would be out of the jurisdiction of what used to be the United States. In the New Democracy of Maine Jacob said goodbye to Sheik Kasan and took a flight to German Prussia that was itself a splinter country from what had used to be a unified Germany.

Suddenly, once on the flight, however, Jacob suffered with a severe case of buyer remorse. Jacob knew from the appearance and demeanor of most of the other passengers on the flight that he was on his way to the Kingdom.

His first assignment was to be an introductory job at a remote post at the encampment at Sala. After making friends, and getting a good report from Sheik Omar Abu Abud he would report to the Technical Center of Khadid for his primary consultation work.

At Sala it was important for him to do routine work of whatever nature was assigned. He was to prove that he could work within the rules of Kamsi society. "It is," as Kasan explained, "a test."

The flight from Prussia was on an old DC-3 that was much older than Jacob. Once in the air the wings vibrated violently until the aging aircraft slowly climbed up to about fifteen thousand feet.

The passengers didn't hesitate to stare at him and made open comments about his white pasty skin. Unlike his years in America when whites would cut their own tongues before making an insensitive comment about others these people were self-assured and in their own element. If this was the appetizer Jacob dreaded the main meal that he knew was coming.

From his cramped position sitting on an orange crate in the cargo hold of yet another ancient DC-3 aircraft Jacob saw one of the other passengers. She had an actual seat and once, when the flight from Khaybar had been especially rough, had gotten up and come back to him.

"Sir, I will sit here on the box if you wish the seat."

"Why no," Jacob replied surprised that a young beautiful woman would offer a seat to a man. "I'm fine, but thank you. Thank you very much."

"My pleasure," she said. Her skin was reddish; her blue eyes told him that she wasn't Arab. He couldn't place her nationality.

"Excuse me," he asked, "What nationality are you?"
"Indian."

She was intriguing. A lovely woman with a sensual quality.

"Are you going to the camp at Sala?"
"Yes. I have a job as a cook," she said.

A cook! Better and better. Jacob had been a bachelor for over sixteen years. Far too long. He had the impression this woman could not only make a man happy, but please his stomach as well. He knew he would have to work fast. No woman like this would stay a cook. Not in Sala where men outnumber the women fifteen to one. He learned her name was Mardi Boland.

"That doesn't sound Indian."

"Oh!" The aircraft bumped again and she sat down heavily in his lap. The orange crate under him buckled and bright bouncy oranges sprang out through the slates and went bouncing along the isle in the bucking aircraft. The oranges seemed to chase one another all the way to the cockpit door.

Jacob braced himself, laughed with Mardi and held her easily with one arm. Suddenly, now with this beautiful woman in his arms he was grateful for lawn mowers with brakes. He had never been in such good physical shape. Mardi weighed maybe one hundred pounds and he cradled her in one arm without effort.

When she stood up her smile was different. She was confident almost commanding. She knew she had made a conquest.

"My name is American. I was born there, but moved to India with my parents when I was a baby. I consider India my home."

She held onto an overhead strap while Jacob slid another orange crate out and got settled again.

"I'm a scientist," he said, "but at Sala I'll work as a technician."

"Why?" she asked.

Jacob started to answer and then cleared his throat. What could he say? That he was on probation for ninety days? That the Kamsi chiefs did not trust him and wanted him to prove himself with some menial work? That his stay at Sala was a condition of his contract?

"Where's your assignment?" he asked.

"Sala," she said.

In a few moments Jacob discovered that Mardi was to work in the house governed by Sheik Omar Abu Abud. They were to be co-workers.

An hour after Mardi went back to her seat the aircraft landed at a dirt airfield twenty kilometers from Sala. Jacob walked down the aircraft ramp carting his heavy suitcase and felt hot air burn his exposed face. His tongue swelled. He had never felt heat so oppressive.

Nothing prepared him for this. Searing desert air blistered his skin on contact. He knew he would have to get out of his western clothes into robes and cover his fair skin especially his face. No wonder Arabs stayed undercover.

On a distant horizon Jacob saw huge sand dunes jutting upward to a deep purple cloudless sky. The dunes visibly moved in a serpentine way casting deep blue shadows over brown sand under relentless pressure from the wind.

Jacob rubbed his eyes and looked again. The dunes actually did move with upper layers of sand sliding down the slopes only to have the top of the dune replenished from the other side. While this was true of all deserts he was astonished to actually see the process working so powerfully. Kasan did say they were having problems controlling the desert and now Jacob saw why.

Trying again to move his tongue Jacob's mouth felt as if it were filled with chunks of hot dirt. Then something else besides desert sands moved as a metallic surface shimmered in the sun, sliding out from the shade of a large tin building.

A large limo motored slowly out toward the aircraft. Mardi, he, four other passengers, and the pilots waited in the stifling heat for their rides to Sala.

Jacob studied the limo. It was a big vehicle, riding high on oversized balloon tires. The limo lumbered toward a security gate that swung open. He watched as the limo motored over sticky asphalt and come to a stop near them.

Jacob couldn't see inside the vehicle because of the usual dark tinted windows. The windows of their vehicles are tinted and their women are veiled. Kamsi, Jacob knew, love their privacy.

"Ebbtide?" a small runt of a man got out of the driver's side. He wore old American Army fatigues. When the door opened Jacob heard girlish laughter inside the limo.

"Yes," Jacob said, but when he stepped forward toward the limo the runt stopped him and offered the open limo door to Mardi.

"Here," the runt said gesturing as he opened a rear door. Jacob tossed his luggage into the back seat and got in. The interior was dark, but spacious. A solid partition divided his compartment from the others up front. Cool air flowed out of side vents. The interior was most inviting.

Behind them a jeep brought the pilots and one other passenger. Well, Jacob thought, at least he had air conditioning.

Laughter from the front seat filtered into the rear. Only the women and he were traveling first class. Jacob shrugged. Maybe the other men weren't going to Sala? He shrugged again. Well, that *was* possible.

Still the temperature had to be over a hundred and twenty and the two pilots and other man rumbling along in the jeep wouldn't last long in the heat.

Thirty minutes later they arrived at Sala with the jeep right behind them. There was no mistaking it. First the ride of the limo changed from a lumbering back and forth motion on sand to a smooth, tight ride on pavement. The jeep rode by them once they got onto pavement. The two white male pilots were draped over in the back seat and looked dead.

Massive houses sat astride a village square. High-whitewashed walls surrounded each house. Water fountains could be seen behind tall wrought iron entry gates. There was as far as Jacob could see no vegetation at all.

The only way to tell where the desert ended and the houses began was the walls. Only satellite dishes stuck conspicuously on flat roofs belied another world out beyond the desert.

As the limo went passed whitewashed houses Jacob saw that most dwellings had courtyards. They were bounded on three sides and open on the side facing the street fenced only with a tall wrought iron fence.

In the middle of the encampment was the village square, another courtyard with the same general layout as all the others except bigger with whitewashed stairs going up to a second floor balcony.

The limo pulled to a stop across the street from the village square. Jacob saw uniformly fat white-robed men sitting on wooden chairs in front of the square eating fruit out of bowls. They lounged casually apparently waiting for something. The limo idled. Jacob leaned back in the seat, his head swayed to one side, studying the architecture of the village square.

Officially, Sala was a Sheikdom. Omar Abu Abdu was the main man. So long as Jacob remained in Sala he was Jacob's employer. While the Kamsi Kingdom had lawyers and corporations its social structure, including its work environment, was tribal.

Sand blew across Jacob's view. A shadow covered the limo's window. When the door opened a heavy-set man wearing cheap plastic sunglasses leaned into the car.

"You are he?"

"I'm Doctor Ebbtide."

"Yes, yes," he said impatiently. "I am Abu Abud."

He took off his sunglasses and looked hard at Jacob.

"I am Sheik Abud," he said again as Jacob slid across the seat.

Jacob extended his hand, but looked again at the small black eyes set in the Sheik's fleshy face. The Sheik did not shake hands. He straightened up, but held the door open politely.

"I am Sheik Abud," he repeated, "but you may use my name Omar."

"Omar," Jacob repeated.

When he got out of the car Jacob first saw water flowing over the lip of a fountain. A driveway gate squeaked open slowly. The hinges needed oil. Behind him the limo door closed and the vehicle started away.

As he turned to watch the limo Jacob's sight fastened on the courtyard across the street. The walls were stark white; and yet, something red stained the interior of two walls as they met at an interior corner.

"Come!" Omar commanded.

Jacob started to walk up the driveway toward the house, but Omar stood in the driveway and shouted something to the men across the street that caused a ripple of laughter and some applause.

Jacob's attention focused on the courtyard interior, especially the walls. Strange, he thought, there's only an upright post in the courtyard. Nothing else except for the chairs and small tables with bowls of fruit.

"Sheik Kasan say you smart," Omar said.

Then in rapid Arabic shouted to the men, who now got up, walked around while watching a small door in the courtyard wall.

A man came through the door under heavy guard. He stifled a cry. He looked across the street and half-walked and let himself be half-dragged to the upright post.

"What is this?" Jacob asked.

"Sheik Kasan say you very smart," Omar repeated, "but play loose to goose."

"Huh?"

"Loose to goose!" Omar insisted said again speaking more to the men across the street than to Jacob.

"You mean fast and loose?" Jacob asked gulping as his gaze now focused on the man having his wrists tied up over his head. The man, naked to the waist, was lashed to the post. Jacob wondered if the situation across the street was somehow staged to intimidate him.

Omar squinted in the bright sunlight and pulled his glasses down over his nose. Walking away, waddling his stubby frame slowly up the driveway, the Sheik stopped momentarily and turned around to watch the man in the courtyard.

Sudden cries from across the street caused Jacob to cringe as sharp lash of a whip whistled through hot air. As the whip fell against the prisoner's flesh Jacob's skin bled water; he sweated worse than with the miserable heat. A cold sweat washed over him and Jacob suddenly wondered if he should have spent his days pushing lawn mowers with brakes.

"We are a people of law," Omar said with his usual impatience. He shouted something in Arabic to those across the street. Jacob stood side by side with Omar. Against the sunlight bouncing off whitewashed walls he saw the man at the post. He wore some type of halter on his head. Jacob pointed.

"What is that?"

"Head brace," Omar said. He put his head down as if bowing with his fat chin tight against his white-robed chest.

"The condemned must not look up," he said. "The brace forces the unrepentant into a proper servitude."

"A proper servitude?" The words surged out of Jacob's lungs with the force of indignation. Jacob lost his balance, stumbled over his suitcase and regained his footing. Reluctantly, he followed Omar up the driveway.

Jacob followed Omar into the house and up to the second floor, but even in the upper hallway the anguished cries of the man at the post echoed loudly off of bare concrete walls.

The door to a room was not locked and when Omar pushed it open Jacob saw a bare whitewashed servant's quarters with a straw mat for a bed. Jacob gave the room the once-over before going inside.

"This will do fine," Jacob said as if to hang onto his last shred of dignity. Walking in he noticed a second mat rolled up and placed neatly in a corner. There was no other furniture and only one small window. It looked out over the village courtyard.

"Perfect," Jacob said while forcing a smile.

"Your man will be here soon," Omar said.

"Man? I didn't request any help. Look, Omar, I'm repairing stereo equipment. Right?"

Omar twisted his face into a look of complete contempt. "But you must have our man," he insisted, "How else can I protect you?" Omar's remark made no sense to Jacob so he tried again.

"Sheik I work alone. I can't afford any helper."

"No," Omar said. "This is your man. Not your helper."

His nostrils flared in sudden anger as a long wail came up the stairwell. The whip fell on the back of the condemned. Jacob shook his head fighting off a sudden terrifying thought.

"My man?" he questioned.

"He stands in your place," Omar said with his hands folded across his ample belly. The flogging stopped, although the moaning of the man was now almost musically soft, mingling with the sounds of water fountains.

"Master," a gentle male voice said.

"Your man," Omar said. "His name is Hasham. You must treat him well!"

"Hasham?" Jacob asked, his gaze going back into the bedroom to the second bedroll on the floor.

"Do we have to sleep in the same room?"

"No Master," Hasham said. The Sheik turned his back on him and faced Jacob.

"You are an infidel," Omar said. "The people will falsely accuse you. Hasham is one of us. The people like him. If they accuse you he must stand at your place at the post. Understand?"

"God!" Jacob raved. "You mean he's my whipping boy? He takes my place at the post if I break your stinking tribal rules?"

"Take care!" Omar said, coming up, pressing his ample stomach against Jacob. Hasham whispered, "Please Master, say not these things. It is forbidden."

"In the west you can steal, rape and kill!" Omar pressed Jacob against the wall. "And, if you tell a judge it was all a big mistake you get away with it! But not here!"

"But, I…" Jacob squeezed his way out into the hallway. "I can't let a man take punishment for me!"

He saw that Hasham was half knelling to Omar and that his 'man' was shaking. "What kind of a man would I be to let another take my punishment?"

177

"Fool!" shouted Omar. "A coward, a weak man would let his man take the punishment!"

"I am a professional man," Jacob insisted. "I have self-respect. I have made mistakes, yes, but I have a good heart."

"Enough," Omar commanded. "Now, yes, Mister, let us say that you break our laws," he turned to Hasham, "He earns over sixty thousand gold dinars a year. Yes. Is it not so?" he asked the man.

Jacob saw a sly smile forming on Hasham's lips. So that was it! Hasham took some chances. He ran interference for foreigners that came to the Kingdom.

The people knew and liked him. They knew that to falsely accuse a foreigner of a wrong Hasham would get the whip instead.

"I see," Jacob said feeling much relieved. "Sixty grand you say?" Not a bad piece of change for a glorified houseboy. Not bad at all Jacob thought. That was damn near as much as he was making as a lead scientist in Israel.

Omar nodded. His eyes squinted against the glare of harsh light off whitewashed walls. Hasham bowed again and went into the bedroom. He picked up his straw mat and walked, submissively, behind Omar down the hall.

Over the next couple days things went well. The house had an old music system that was to be replaced, although Jacob knew that in strict Kamsi doctrine music is completely forbidden.

Still, Kamsi society was tribal and there were differences in what was considered acceptable; and, in Sala under the rule of Sheik Omar Abu Abud music and good food flowed like the River Nile. In a day or two after the initial shock of having a 'whipping boy' assigned to him, Jacob found that he was enjoying his stay.

There was virtually no crime. Payment for any service was paid in gold not worthless paper currency; and, after his first day he knew that the only real use for the courtyard across the street was for the old men to sit around and eat and tell lies to one another.

Jacob started on the first floor. He checked wiring and stereo equipment signals. He worked alone and made good progress. The house was to have music 'piped' into every room except his.

That his room would not have the dubious benefit of music was either an oversight or an insult and Jacob didn't care one way or the other. As he worked he never saw Hasham, but when he went to the large kitchen for his meals Mardi was always there and always smiling. Jacob ate well. After only a week his stomach ballooned so that he had to loosen his belt.

Mardi waited on him and seemed to enjoy his company. One day, as Jacob sat in his chair, patting his contented belly and grinning boyishly up at Mardi she leaned closer to him.

"There's nothing to do here," she complained. "I'm bored."
"Is it a problem?" Jacob asked. "Take each day as it comes. Things change."

"I'm bored," Mardi pouted. "I need to find something else. I'm sick of Omar. I'm sick of this house." She cast a sour glance at the kitchen. "I'm sick of pots and pans."

"Do you have a man?" Jacob asked. Mardi straightened up, cocking her head to one side studying him.
"Why do you ask?"

Jacob laughed. "Oh no, you misunderstand. I have a man, see?" he said. He explained about Hasham.
"He's my insurance policy."

"You must be very important," Mardi sat down next him, putting her elbows on the table and her chin in her upturned hands. "You are a VIP."

"A VIP?" Jacob asked grinning. "Where did you pick that up?"

"In India? I lived there, but learned English. I watch television."

When she got up Jacob was worried that the one sane person in the house was going to leave. Mardi went to a storage area behind the kitchen, but turned to Jacob.

"I have a bag to lift," she said. Her tone of voice was businesslike. "Could you?"

"Sure," Jacob said. He followed her inside the storage room filled with shelves and bags of vegetables, but as he stepped inside Mardi turned to him and threw her arms his neck kissing him hard on the mouth.

Jacob stood in the doorway both pleased and startled. He made no move. His arms were flat against his sides as Mardi pressed her mouth onto his.

"I'm bored," she said again. "All the men here are fat and wear big black boots."

When she took a step back Jacob didn't follow, but simply stood in the doorway watching her. Jacob smiled and let his imagination wander. His infatuation grew as his gaze wandered away from her lovely eyes to other areas of interest.

Mardi breathed deeply, her breasts moved in and out through a light fabric blouse. Jacob started to say something, but she brought her fingers up to his mouth.

"Shish," she whispered coming closer again.

Jacob moved his head slightly. With any woman it's never good to appear too anxious or too willing.

"You're right," he said. "The men are fat and wear black boots and carry rifles. They treat their women like mules."

Mardi's eyes sparkled in righteous anger. "Like mules," she whispered.

"Like mules," Jacob whispered again.

"Worse than mules," Mardi said softly bringing her lips to his.

When she pulled him down he couldn't resist. They were on the cold stone floor making love. Jacob knew his hunch about her was right. Mardi would make him very happy. He murmured a quick marriage proposal to her, but she giggled and whispered, "It is not your head talking my love."

"Mardi," Jacob said stroking her long braided hair,

"I love you," but quickly felt a sudden emotional rush of dread as a strangely familiar voice shouted, "Mardi!"

She froze under him and then pushed him off her. Hasham stood looking down on them lying in the pantry doorway. Hasham's face was distorted with hate.

"Woman!" he whispered harshly, "Unfaithful bitch!"

11

Mardi was on her feet. Her concentration on Hasham.

"I told you," she said defiantly, "I want a divorce. I married you when I was thirteen and…"

"You're married?" Jacob's voice squeaked.

He pushed himself up to his knees and stayed there as Hasham took Mardi by her shoulders shaking her.

"You do this to get me under the whip!" he shouted. "Unfaithful bitch!"

Hasham's slap would have cost him ninety days in a Remnant jail for spouse abuse. Mardi pulled roughly away from him. Hasham put his hands on his hips and stared down at Jacob who was unable to find the strength to stand up. Jacob gulped and raised his hands beseechingly. Hasham reached down yanking Jacob to his feet.

"It is the law here," Hasham said. "A man must control his wife! Women are weak. Men are to be strong! That is the Code!"

"My god Hasham," Jacob mumbled. "I had no idea!"

Hasham raised his head as a door slammed someplace in the kitchen. Mardi raised a finger to her lips for silence, but it was too late. Omar and a guard strode into the kitchen. Omar stared from Mardi to Jacob and to Hasham.

"Adultery!" admonished Omar glaring at Jacob.

"I didn't know," Jacob pleaded. "I was set up. The bitch set me up! Honest to God, Omar. The bitch set me up!"

"Ignorance is no excuse!" Omar shouted. "In America one can rape and kill and commit heinous acts and go free, but not here!"

"That isn't…" but then Jacob caught himself. This was no time to argue for a rule of law. These thugs were going to whip somebody. He cringed when the Sheik, the guard, Hasham and Mardi, all four of them, looked hard at him.

Omar said something to the guard in Arabic. In five seconds or less he, the guard and Hasham had left the kitchen. Only Mardi and Jacob remained.

Jacob felt a need to be alone. He brushed roughly passed her. Mardi said something, but in his haste and confusion he didn't hear. He went directly to his room feeling like a child caught dipping into the cookie jar. An hour went by when a knock came to the door. Omar stood there.

"You do not work?"

"Ah, yes," Jacob said. He wiped sweat out of his face. "I needed, ah." He felt forlorn and more than a little scared.

"We are not fools," Omar said. His manner was polite and not at all threatening. When Jacob took a large breath of air nodding his understanding Omar came closer.

"You play loose to goose. Here you must follow rules. Is good that Sheik Kasan sent you to me for training. Yes?"

Jacob had to admit it. Kasan had him figured. Yeah, as Omar said he did play, 'loose to goose.' It was as good a way of saying it as any. Another two and half months of 'training' and he'd be savvy enough to survive in Kamsi society.

"Yes," Jacob replied, "I play loose the goose, but that is now history. Now I follow the rules." Omar seemed pleased and Jacob wondered again about his whipping boy and not about Mardi. She was history.

Was it all a setup to scare him into subservience? When Omar left him standing in the hallway Jacob realized that, even if it was all an act, it worked. He resolved to play by Kamsi rules.

His options had run out. He was a criminal in the Remnant and Leviman would probably have him shot if he went back to Israel. *That* situation still puzzled him.

In the following days and weeks he saw Mardi only briefly. An open door here or her figure going passed a window. Nor did Hasham offer any conversation.

His 'man' did what he had to do and mostly that was nothing. Jacob's 'insurance policy' went somewhere during the day and appeared briefly at mealtime. He showed up with food and sometimes with a snack of fresh fruit.

Hasham sat the food down on a small table and without speaking walked away. Jacob felt that Hasham didn't like him, or maybe he didn't like Jews.

Once the music system was installed in the house Jacob was called on to install two satellite dishes and to install re-transmitters to send signals to television sets to any room in the house. Time, as they say, flew. After hours one day, when that job was mostly finished, Omar stopped Jacob as he walked through the servant's quarters to his room.

"I talked with Sheik Kasan," Omar said. He came into the bedroom. "You have metal in your feet. That might cause cancer."

Jacob had forgotten about the metallic pellets under the skin of his feet. He still felt them rolling under his skin, but all the walking pushing lawnmowers had strengthen his muscles, even his feet.

"Yes," Jacob said, "I'm concerned about that."

"Sheik Kasan thinks you are ready for your assignment, but first to get the metal out of your feet."
"Thank you," Jacob said.

For the first time Jacob saw Omar smile. He seemed happy. Maybe that Jacob was moving on to his permanent assignment gave the Sheik some satisfaction. Jacob's 'graduation' would make Omar look good to his superiors. It made Jacob feel good too.

"You will go to the hospital at Al Serval," Omar said.
"They have fine doctors, but remember," his expression clouded over, "You may still be tested. Be on your guard!"
"I will," Jacob agreed.

"They are Kamsi! They do not know you. Be careful!"
"Yes," Jacob said smiling. This time Omar shook Jacob's outstretched hand.

"Thanks. Much appreciated," Jacob said.
Two days later Jacob was admitted to the hospital and surgery was done the following day. Things, however, did not go well.

"They have quite a sense of humor don't they?" a man's voice asked. Jacob heard, but didn't respond, because his mouth was too dry to speak.

His eyes were swollen shut. He felt the bed sheets and tried to relax cramping leg muscles. He remembered entering the hospital. He remembered the argument between Omar and the doctors, who insisted that removing the pellets from the feet of a wanted felon would cost the hospital needed money.

Omar ushered the doctors into a small lounge area and closed the door. Jacob had listened. "He is a scientist," Omar said. Strangely, for some reason both men spoke English, although, obviously they would have been more comfortable speaking Arabic.

"He's wanted!" said a Doctor. "In America."
"He has knowledge of the Solbean plant. He is here to help us. We must help him!"

"Solbean?" asked one of the Doctors. "What is that?"

Omar then spoke rapidly in Arabic, his voice rising, and then in English, "We need him! The surgery is necessary!"

After much grumbling by the hospital staff and more insistence from Omar the office door opened and the doctors filed out. The surgery was approved.

Jacob remembered the hospital room. His feet were raised up on a board and the surgeons stood around looking at the bottom of his feet and probed here and there with a steel knife. Then he had been put to sleep.

When he woke up his mouth felt filled with plaster, his eyes were swollen closed, and his feet were raw and burned. He gently nudged one foot against the sole of the other. Jacob thought the pellets were gone, but he wasn't sure. Something, however, had certainly gone wrong in the surgery. Jacob heard the man moving around by his bed.

"You'll be fine," the soothing voice said.

Jacob decided the speaker was a refugee from the United Remnant. A hireling like him. Jacob wondered idly if this guy ever had metallic pellets put into his feet and was ever subjected to his feet burning from the inside out?

"I'm Raymond Rey," the man said. "I work for Omar." From his tone of voice Jacob decided that Mr. Rey was a self-satisfied man.

When I get out of this bed, Jacob thought, I'll have the sweet satisfaction of strangling this guy with my own two bleeding hands.

"Did you catch my name?" he asked.

"Rey," Jacob mumbled. "Yeah. I'm good with names."

"Don't think of revenge," Rey said.

Jacob slowly balled his fingers up into a fist. His hands were itching to wrap themselves around Rey's neck.

"Revenge is a waste of time," Rey said. "People have tried that. They get hacked to pieces for their trouble. They have a rather stern Code of conduct over here not at all like the U.R."

It wasn't so much what Rey was saying that aggravated Jacob it was his sickly sweet, ever so civilized, tone of voice. If ever a bastard needed a good thrashing this bastard was it.

This guy would go first Jacob resolved. If there was anything he was coming to hate worse than a small-minded Kamsi it was sanctimonious muck-mouthed U.R. hirelings. First, this bum would pay and then the doctors in this hellhole would be next and their code of conduct be damned.

Jacob breathed deeply through his nose. The air was cool and moist. He found it refreshing. His skin seemed to have some lotion or salve on it.

His skin was cool, except for his feet. A straw touched his lips. Cool liquid flowed in icy streams around his parched tongue. The skin of his mouth absorbed the liquid before it reached the back of his throat.

"Here," Rey said.

Yes, Jacob thought. He'd kill these damn doctors, but whoever was decent enough to give him a sip of water would be spared. Maybe Rey was all right, but he'd make a list. When in Rome thumbs up and down.

"When will he be ready?" the voice belonged to Omar.

"The nurse," Rey replied. "She has been disciplined."

"But his face is...?" Jacob listened as Omar seemed at a loss for words.

"Carbolic acid to sterilize his feet was thrown into his face," Rey said."

"How?" Omar shouted.

"The nurse. The nurse," Rey insisted. "His bandages will come off tomorrow. Now he needs rest."

"So be it," Omar said. "You come with me!"

The Sheik did not seem pleased with his hireling. Jacob heard the door close behind them. So, Jacob thought, that's what happened. They bathed his feet to clean them and then threw the acid bath in his face.

Now the nurse, hmm, yes, he needed the name of that damn nurse. She was now on the top of his hit list. Jacob sensed through his bandages that the day was nearly over. The sounds outside his room died away taking on a quality of night. The rhythms slowed. When people spoke their voices were more hushed and finally, after a long while, Jacob went to sleep.

He slept fitfully, because his bandaged head ached with first-degree burns. The bandages were very uncomfortable. The bandages had a thick outer skin of gauze; and, in between the folds of cloth, some sort of smelly ointment had been applied in a thick creamy layer.

At times in the night Jacob thought he heard the door open with an accompanying rustling sound by his bed. Then a dream started. A dream voice, low and trembling, kept waking him up. Over and over a plaintive man spoke. Jacob wasn't sure, but it sounded like Hasham. Jacob heard him pleading. His voice was low and somber.

The dream came and went fitfully as Jacob slept and he seemed to be begging that something not happen. When Jacob turned over, or woke up, raising his head off the pillow, the dream stopped.

Then, after a few minutes, and Jacob's breathing evened out in sleep again, the dream returned. It went that way through the night.

"Do not harm your loyal servant Hasham," he said when Jacob woke up in the morning. "I am ready to do your bidding, Master. Do not harm me."

"Hasham?"

"Yes Master."

"Harm you?"

"Yes Master."

"I won't harm you," Jacob said.

That morning the bandages came off Jacob's eyes. He was sitting up in bed. As a swath of gray cloth moved away from his eyes he saw a hand and then a blurred image and then two of people: A man and a woman, both Kamsi, with serious expressions, unsmiling. The woman was a nurse. She reached for another part of the bandage and tore it off roughly off Jacob's reddened skin.

Jacob said nothing, but Hasham reacted, grabbing her. The nurse pulled away from him and then reached again for another bandage as the man, a doctor, looked on indifferently. Jacob held up his hand.

"I'll do it," he said.

"Not allow. Not allow," the doctor said. "The nurse will take off bandage."

"No she won't."

Jacob motioned them back from the bed. His skin was tender but seemed to be healing nicely. The rest of the bandages slid off without tearing skin. Jacob noticed a small patch of blood on his hands from where the nurse had ripped off the first piece of tape.

The nurse came forward once, reaching, but Jacob's hand came up, warning her away from him. Both the nurse and the doctor kept their distance, sulking.

One by one Jacob took the tape and bandages off his head and then started unwrapping his feet. Jacob pulled his feet up to him in the bed to inspect the surgery.

A guard appeared in the hallway. The doctor said something in Arabic. The guard took a rifle off his shoulder. He came into the room with the weapon pointed at Jacob.

Jacob forgot about the guard, grabbed the doctor's hand, pulled him violently over the bed and hit the doctor on the nose with the palm of his other hand. Behind the physician Hasham made a noise. The doctor bounced off the bed.

Hasham tried to catch the man as he stumbled to the floor dropping a pair of scissors. The doctor cupped his hands over his bloody nose. Hasham hurriedly said something in Arabic as the doctor got up on his knees. He had the scissors cupped in both hands. His eyes were hard on Jacob.

"No!" Hasham shouted in English.

Jacob blinked and the doctor was on the floor gingerly exploring the contours of his bloody broken nose. As Jacob swung his feet off the bed, taking his weight on his tender feet, the nurse tripped over the doctor on her way around to the guard. In an instant the nurse grabbed the guard's rifle.

"We leave now!" Hasham yelled.

Jacob saw the nurse get possession of the rifle. She swung it around pointing the barrel at Jacob's chest. It clicked. The nurse grimaced, her eyes distended, her nostrils flared as she fumbled with the rifle. She and the guard shouted hurriedly in Arabic as the guard rushed forward again to get his weapon. They fought for the rifle, but almost let it drop to the floor. The nurse yelled something as the guard let go the rifle. She swung the weapon around.

Jacob wondered, watching them, why the guard didn't grab the weapon. Instead, he seemed content to let the nurse have it. Hasham was the first to react.

"The gun!" he shouted hysterically in English.

Hasham stepped on the doctor's chest in an effort to get the rifle away from the nurse. He reached for the nurse that managed once again to squeeze the trigger. Again the weapon failed to fire. For an instant Jacob had the insane notion that maybe the weapon was not loaded?

Jacob blinked as the gun apparently jammed. The nurse shouted in anger as Hasham lunged for her. Her eyes flashed. Both Hasham and the nurse yelled at each other in Arabic.

Hasham reached her and fought for control of the rifle while Jacob and the guard stood immobilized staring at each other.

Jacob sat down on the bed wide-eyed as the guard belatedly sprang into action. He rushed around Hasham and the nurse, stumbled over the doctor and lunged for Jacob.

Years before Jacob would have been idealistic enough to try to reason with women and children with guns, but now he jumped up and ran around the bed and gouged at the kid's eyes and threw the guard roughly onto the bed as the nurse wrestled her way clear of Hasham.

She still had the rifle! As the nurse ran around the other side of the bed yelling in Arabic she brought the rifle up to her shoulder again.

Hasham was separated from the nurse by the doctor who insisted on getting up by grabbing hold of Hasham for support. Hasham hit him on the nose again. As the doctor yelled he grabbed hold of Hasham's mid-section and hung on. Hasham dragged the doctor as a dead weight around the bed as he tried to reach the nurse.

It was obvious to Jacob, however, that the nurse was not used to handling weapons, because she held the rifle out in both hands, trembling as she struggled to control the heavy weapon. His man punched the Doctor again.

Hasham and the Doctor struggled, clutching one another, trading blows. They grabbed one another, hanging onto each other, like two punch-drunk prizefighters.

They lurched around the bed in the direction of the nurse. As they went by they pushed heavily against Jacob, who still standing by the bed turned around and around trying to keep the action in his line of sight. The nurse, the doctor and Hasham seemed to be waltzing in slow motion, grabbing at one another and all the while yelling in Arabic.

The rifle came back close to Jacob's head as the nurse swung the rifle around almost handing it to Jacob. Grabbing the rifle, pulling hard and finally getting his sore feet take his weight Jacob kept his strong hands tight on the rifle. The nurse was no match for him.

He had the nurse off-balance. She tumbled onto the bed. While the nurse collapsed face-first a noise at the doorway distracted Jacob. Other guards rushed into the room.

Jacob fell on the prone nurse, bringing the butt of the rifle down on her head. As the rifle stock struck her head she bounced on the bed, tumbled off and landed with a thud on top of the doctor that was on the floor again.

It seemed like a lover-tap as the guards rushed at him. Jacob shook his head, cleared his vision, swung the rifle around, aiming it down at the nurse, but the next instant he woke up.

Blackness floated away from him as syrupy oil draining into a bucket. Eyes blinking, head throbbing, Jacob's neck felt funny. Jacob tried to look around, but his neck lacked the strength to raise his head. He was dazed and could do nothing but stare down at stained concrete. Strong hands stopped him from falling.

Memories floated in and out of consciousness. Was he in the hospital? In another room? Yes, he remembered. Had there been some sort of quick tribunal? Jacob remembered standing on a richly ornate Persian rug. Questions were asked. Charges had been made. Hasham had spent the whole time on his knees, hadn't he?

"Doctor?" Jacob asked. "Are you a doctor?"

Two men held him up. Neither made reply. Sunlight pushed back a shadow. Jacob realized he was outdoors. Swinging his head side to side he saw a courtyard. White columns marched around in stately style while doors and windows opened onto the courtyard from setback corridors.

Faces peered out of deep shadow. Jacob focused his eyes on a small group of white-robed men standing next to a doorway. One of them looked at him.

There was something terribly wrong with his neck and head; he struggled to orient himself. Then one of the guards, who held him upright, released him.

Weak-kneed, Jacob collapsed. One guard walked over to a white-robed man. He whispered something to the guard. The guard walked back and took hold of Jacob's arm again.

White-robbed men moved into the shade of the colonnade. Jacob tried to clear his head by concentrating on objects around him. First, the general layout of the courtyard came hazily into view. It was barren concrete. Then, by the stairs leading up to the second level, he saw the staunch wooden whipping post.

He stared at the post and a huge stone slab, gray and flat, at its base. His heart beat faster. Jacob felt a mixture of fear and contempt. The guards half led, half carried him in the direction of the post. Jacob's head was groggy, but he was suddenly irritated by the sudden knowledge that he was wearing the head brace. His head hung over in an attitude of abject humility.

It especially galled him to know that he looked like a remorseful criminal led to a just fate. He found himself masochistically wondering if they hacked off fingers or toes first or got right down to serious drawing and quartering.

Approaching the group, his eyes hurt painfully as the sun edged into the courtyard, firing silver glints of sharp light off flicks off gemlike stones embedded in rough concrete.

Jacob panicked. He fought hard to control himself. His mind raced through any last desperate avenues of escape, from loudly asserting his UR and Israeli citizenship, which was always good for a laugh at a hanging or other shindig, to begging forgiveness of the nurse whose bits of skull he had scattered around the hospital room floor.

More than anything else Jacob wished that blasted head brace was off. At least he could die looking this collection of fat Kamsi in the face.

He resented dying like an old horse lead to a glue factory. He had to do something. Anything! When one of the guards moved around in front of him Jacob put his weight on one foot and stomped down hard on the guard's foot with the idea of breaking free and making a run for it.

But when his foot came down a terrible pain seared upward through his leg. Through sudden tears Jacob focused his wavering vision down to his right sheepskin boot.

Each of the guard's boots had a small nail sticking straight up on the metallic toe of the boot. His tender foot was impaled on the guard's boot. Rearing his head back against halter Jacob bit his tongue. He mustered all his strength to stifle his screams.

Through his pain Jacob heard ripples of laughter. Even the guard thought it was funny. Reaching down the guard grabbed Jacob's leg and slowly, bending Jacob's leg at the knee forced Jacob's stuck foot free of the upturned nail. Jacob howled and collapsed into the guard's arms.

His wounded foot felt as if he had danced a jig on the electric carpet. Now he was a lame and haltered horse going to the glue factory.

"What have you to say?" one of the white-robed men asked as the laughter died away.

Jacob's foot was on fire with pain. When he wiggled his toes he felt hot sticky blood sloshing around in the animal hide boot. Cries of pain and outrage gagged in his throat.

Trying to raise his head was useless so Jacob tried to look up at the man from under his head brace, but even that was no use. Jacob kept his teeth clinched tight against the pain and the indignity of breaking down completely.

A long moment passed with nothing by his staggered breathing, suppressed grunts of pain and tears rolling down his cheeks. His tears embarrassed the hell out of him. Jacob prayed to God for freedom and a rifle.

Every cell of his body regretted letting those guards get behind him. He could have saved the nurse for last: She was prone on the bed and he had the goddamn rifle. He could have settled with her anytime.

The over-ripe Kamsi said nothing, but when sufficient time had elapsed he clapped his hands twice. A door opened immediately behind Jacob. He heard a shuffling as if somebody was being dragged. A vaguely familiar voice begged for mercy.

"Please Master!"

Where had he heard that voice before? Where was Omar when he needed him? By turning his head sideways Jacob saw the thin, dark man and memories flooded back.

"Hasham!"

The guards dragged him to the post, but he was fighting them heroically. "No," he yelled. "I have done nothing! It wasn't my fault!"

A fat Kamsi came up confronting Jacob. "Well?" he asked and started to speak again when Hasham, with a strength borne of desperation, lunged away from his captors. He got six paces before the guards captured him again. They dragged him yelling and struggling to the same spot in the courtyard.

Suddenly Hasham broke free of his captors again and rushed to Jacob, bending down to Jacob's wounded leg, grabbing it in desperation. Hasham's hands wrapped themselves hard squeezing already cramped muscles. Jacob fell over on Hasham with a mixture of sobs and grunts.

"My family needs me!" Hasham cried as the guards pulled him away.

Jacob strained in the halter, hopping on his one good foot, screaming in pain. Through his tears Jacob saw Hasham's face and hands coming free of his boot covered with blood.

"I…I don't understand!" Jacob shouted.

"He is your man," the fat Kamsi said. He said it indifferently as if quoting the price of camels take it or leave it. The Kamsi clapped his hands once and Hasham shut up in the middle of a sob. He knelt motionless. His tears fell onto his clothing. One instant he was begging on his knees and the next the wind blew over the courtyard; the silence was complete except for Jacob's jagged grunts. Before Jacob could think of anything to say the Kamsi took a few steps forward.

"Death is the penalty. Begin."

"No!" Jacob shouted. "Don't!"

Again the fat Kamsi walked up. "You take his place?"

"The nurse attacked me," Jacob said making an effort to keep his voice normal. "She attacked me."

"She hurt you?"

"What?" Jacob leaned up against his captors and let them take his weight. At the post Hasham was making sounds again. Low murmuring sounds.

"She hurt you?" the Kamsi repeated.

"Burned my face with foot wash," Jacob said. Now they were getting someplace. This was more like it. They were going to establish the facts. That whole situation had been the fault of that nurse. She caused it.

The Kamsi leaned down, stooping. Jacob's eyes focused on Omar. He held his gaze for a long moment. The Sheik knelt down his expression stern, but also sad.

"Your face is not burned Doctor Ebbtide."

"What?" Again Jacob's felt his strength waning. God, no, he moaned. "Omar," he begged, "Please."

"The gun had no bullets Doctor," Omar said his voice now betraying emotion. "You failed again!"

"But the rifle!" Jacob shouted.

"Do you really think we would let you get shot in our hospital? What?" Omar shouted. "This is not America!"

Jacob's tired mind flashed back to the hospital room. The guard had stood motionless with that stupid smile. The nurse aimed the rifle. The guard knew there were no bullets in his weapon!

"Omar I swear," Jacob whispered while the Sheik knelt down looking at him, "I acted in self-defense."

"You attacked the doctor and the nurse. They are both hurt. You are a guest here. If a foreigner attacks a Kamsi the penalty is death."

Omar stood up. Jacob could only see his hands clasped across his belly. "You should have been smarter. Kasan said that you play loose to goose!"

"Omar, please..."

"You or Hasham," Omar said his voice firm.

Jacob managed to stand up taking all his weight on his left foot. His brain alternated from blank to thinking that 'whipping boy' was some sort of ruse, that maybe... Then, suddenly, in his desperation, Jacob understood. Ah ha! So that was it! This was all part of the test!

"What a fool I've been!" he shouted.

A ripple of pleased acknowledgment circulated through the collection of white-robed men. At the post Hasham collapsed, sobbing, beginning a wailing prayer. Jacob caught his breath. Why of course. That was it! It was all an act!

Omar came up close. "You must choose now!"

Jacob's mind whirled. In spite of his injured foot, his stiff neck and sore muscles, he could hardly contain a wild peel of laughter. Such was his relief!

He tried to nod feebly, to congratulate Hasham on a truly fine performance, but again his tongue stuck to the roof of his mouth. Jacob felt like laughing and yet tears rolled down his cheeks. He was suddenly very cold and yet sweat poured out of him. Omar grunted.

Hasham was calm now. Rarely had Jacob been so moved by the sincerity of an actor's performance. "That man deserves a raise!" Jacob shouted, but of course these Kamsi can afford the best.

Jacob was weak, but still had to give credit. What performers! These Kamsi really knew how to put on a show. They continued right up to the final scene.

Another peel of laughter and relief died in Jacob's throat. Hasham cried out in a terrible wailing. Jacob would be glad when the performance was over.

Swinging his head around in the halter Jacob saw Hasham at the post. Something was wrong. Men were tearing his clothes off, lashing his hands up to the post. Hasham hung with his toes barely touching the stone slab.

As the whip fell, Jacob yelled, hopping on his one good foot. The whip fell again and Hasham was limp at the post. Then Jacob heard the outraged shouts of some stranger, someone removed from himself.

Then as suddenly as the whipping started, it stopped. Hasham was unlashed from the post and placed prone on the huge stone slab while a hooded man with an axe came forward, standing over the prostate man and then, as Jacob's gaze followed the axe it fell in a wide overhead arc. Jacob mercifully lost consciousness.

On awaking Jacob saw that he was back in the hospital. Omar stood by his bed. When Jacob looked in his direction Omar took a piece of paper and read a statement from it.

"Finish your tour of duty faithfully and without further protest or the whipping post will be your companion in the final moments of your life!" Jacob was astonished at the change in the man.

Two days later, sitting in a Centas seat on his way to Khadid, the primary science center for the Kamsi Kingdom, Jacob tried sleepily to put his situation into perspective.

True, he had signed the contract. True he had even seen the small print about the 'whipping boy.' Although he had learned somewhere that Dark Age princes had whipping boys, companions, who would take a flogging if the young prince did something wrong, Jacob hadn't thought of himself as a prince, nor the Kamsi Kingdom as a Dark Age realm.

What a mistake! Why, he wondered, didn't he take them at their word? They had never lied had they? *Why hadn't he taken them at their word?*

Now, here he was, on a Centas, half bus and half railroad train rolling over the desert on massive balloon tires. Each car had electric drive wheels and articulated steering that rolled the machine over the sand like a huge centipede.

"Snack?"

A veiled stewardess offered a selection of small straw baskets filled with fruit. Jacob gratefully accepted. He found a fresh pear in the fruit basket. It melted in his mouth. He munched fruit savoring the simple pleasure of food and life.

Jacob sat watching the desert go by and ate. The fruit was delicious, but after a dozen or so mouthfuls he couldn't enjoy it. The seat was deep leather and yet he couldn't relax. He was stiff and ill at ease.

Maybe even someday finding some way to make things right with Hasham's two children. Mardi had sent him a note telling him that she and the children were going back to India. He had thrown away the note and with it her address.

The last thing he needed now was to get involved with Mardi and the inevitable accusations that he and she had 'cooked up' that confrontation in the hospital to kill off her husband the erstwhile whipping boy.

Maybe he could serve out his contract in the Kingdom, but then what? He cringed realizing that Hasham's kids were not the only problems. He was wanted for multiple crimes in the UR. Flight to avoid prosecution and a jailbreak. Not to mention a long list of felonies for sabotaging those satellites. And what about Leviman?

Jacob finished the fruit and put the basket down on the seat next to him. His head nodded over sleepily. His stomach was full and, in spite of his troubles and anxieties, the deep comfortable seat worked its magic on him.

12

Jacob slept, but in his sleep he slowly became aware of another presence. The gentle rolling of the machine made it difficult for him to wake up.

"May I join you?" The speaker was a tall man. He was well dressed in a western style business suit.

"Name is Raymond Rey," he said extending his hand.

Jacob ignored his outstretched hand, rubbed his face, and looked to the passenger cabin. He remembered where he was, and he remembered the voice from the hospital.

"May I join you?" Rey asked again.

"Sure," Jacob said sourly.

Rey sat in the seat facing Jacob. "The mind adapts quickly," he said. "Are you feeling better?"

Jacob closed his eyes. "Is this another test?"

"No," Rey said. "I'm here on assignment from Omar. I must say that he is not pleased with me. He sent me to the hospital to make sure about you. You know that things were routine. My timing was bad. He and I walked out before," he hesitated adding,

"Omar's worried about you." In response to Jacob's grunt of disgust he added, "If you don't finish your assignment it will reflect very badly on him."

"Breaks my heart."

"Omar is more concerned about your head than your heart. You have no protection now. He's very concerned that you'll lose your temper."

Jacob thought about it. Yes. Omar had reason to worry, but how in hell was he to know that the rifle had no bullets?

That the whole fuss between him and the nurse had been a ruse, a test? Jacob rubbed his face again feeling regretful. Yes, damn it, he should have realized that they were up to something.

"Tell Omar I'm sorry."

"He knows," Rey said, "but still he's worried."

Rey leaned forward looking intently at Jacob.

"What is it?" Jacob asked. "I know you want more than chitchat."

"A delicate subject," Rey said. "Omar wishes to give some advice, but is also worried about your reaction."

Jacob threw up his hands. "So say," he said, "Tell me what worries Omar."

"You situation is doubly bad," Rey said, "Because you are a Western man and also because you are Jewish."

"I can't help either one!" Jacob snapped back. "What am I supposed to do about it?"

"See?" Rey asked, "Omar has the impression that Jews have, for want of a better expression, a bunker mentality. Is it true that Jews believe everybody is against them?"

"Everybody is," Jacob replied. "We are a people apart."

"That attitude will get you killed here," Rey said.

"Try to fit in. The Kingdom needs you. You can help provide energy to a desperate world," and he pointed upward, "but it's all controlled by satellites.

"If we lose the satellites the whole harvesting operation, all of it, becomes chaotic."

Jacob put his head back in the seat. "I'll try to meet them half way," he agreed. "Omar has reasons to worry. Hell, I'm worried."

When Rey said nothing Jacob closed his eyes again thinking. He wondered if the Chinese still used the same materials the same satellite technology as the Americans. It had been those satellites that Jacob had 'penetrated' to take control away from the Americans—he knew he could master that technology, but what if the Kamsi, in an effort to get free of the Americans, had contracted with Russia?

Jacob knew that his useful life as a scientist, with the skills to wrestle control of satellites, was limited. The technology kept changing and the Russians were especially adept at developing software codes that were very hard to crack.

His hunch that he had a short 'useful life span' had less to do with getting killed and much more to do with his realization that his knowledge was fast becoming obsolete.

What to do? He couldn't go back to the United Remnant, because Gladys was waiting for him. He was sure that, sooner or later, government agents would get their hands on him. Now the Kamsi Kingdom had recruited him.

Jew or not he needed a place to live. He needed a place where he was not a criminal. He was in the Kingdom and he realized had to make the best of it.

"You should have read the contract," Rey said, changing the subject. The subject of Hasham grated on Jacob's raw nerves.

"I did read it!" Jacob said. Heads in the passenger compartment bobbed around.

"Yes, I suppose. Still, most people don't have the kinds of problems you've had," Rey insisted. "Many like working here."

Jacob decided he didn't like this man and the best approach to this was to stare forward to end the conversation, but Rey seemed anxious to talk.

"Look," he said, "I think I know what's troubling you."

"You're a mind reader."

"No I'm not, but I know Americans. Generally, they don't know that whipping boys are very highly paid and…"

"Wonderful!" Jacob said. "I hope it was worth it to him!"

"And," Rey continued politely, "And, they take out insurance policies. Don't you see?" He paused, getting a sense of Jacob's reaction, and adding, "To them it's a job. Risky sure, but Hasham signed a contract the same as you did. You have an insurance policy if you die in the Kingdom don't you?"

Jacob shook his head. No, he hadn't any insurance, nor did he realize that insurance was offered. "Never knew," he said.

"Oh well you should have insurance."

Jacob nodded irritably, although he didn't see what his job, or his insurance policy could possibly have in common with a poor guy having a butcher hack him into four sections. Again, as if sensing his thinking, Rey leaned closer.

"Sure they take a risk. They risk their lives for a mere sixty thousand gold dinars a year. All they do for the money is lay around and eat and enjoy the pool until contract time rolls around. And they all sign up again because, you see, the odds are against getting a sore-head loser like you."

Jacob swallowed hard. Tightness behind his eyes told him he was getting angry again. To ward off an outburst like he'd had in the hospital he decided to force himself to be calm.

"So? He had insurance. So what?" Jacob asked.

"Hasham carried two millions gold dinars of life insurance payable if he was killed in the line of service, which of course he was."

When Rey stopped talking Jacob looked over. Two million in insurance? He doubted that Hasham had considered it worth it at the end, but still for someone with limited job skills sixty thousand gold dinars a year was good money. Two million was even better.

"Nothing can make up for the way he died," Jacob said."

"True, but he took the risk. Nobody forced him."

"Tell Omar thanks," Jacob said conveying both his sadness and anger at Kamsi justice. "Tell Omar that I appreciate the information."

"Hasham was a good man. A friend of mine, a Frenchman, had him for a couple of years, but as I say try not to feel too badly about it. Concentrate now on serving out your contract."

When Rey got up and walked out of the compartment Jacob stared out the window. He resolved to finish out his tour of duty and to keep his temper in check. Maybe with some luck he could make it through. Still, he couldn't help thinking that without the trouble in the hospital Hasham would still be collecting his money.

The reality that his whipping boy had been executed sank into Jacob's thinking. He shrunk lower in the seat. Jacob prayed for the first time in years as he realized that any further outbursts would mean a close encounter with a butcher's axe.

While he was mulling over his odds of survival in Kamsi society Jacob wondered again about getting to Israel. He wondered about trying to square things away with Leviman and making a fresh start.

Maybe he could get his job back at the University. The thought of making a fresh start in Israel relaxed him. A sense of peace filled him.

The stewardess came by his seat at dusk and turned his seat down into a bed. A curtain was strung across a railing above his head. Jacob stripped to his skivvies and slept through the night.

The lumbering motion of the Centas, the quiet hum of the motors and the babbling of other passengers gave him a good's sleep. He had forgotten what a good night rest was.

In the morning the Centas moved onto a steel railroad track. The ride was stiffer. The speed faster. After a couple hours on the rails the desolation of endless sand gave way to modern buildings. The Centas stopped short of a massive dome structure, a huge domed building, apparently waiting for a dispatcher to direct it onto a rail siding.

The wait lasted for an hour. Then the Centas entered the terminal. Jacob saw train after train lined up on parallel tracks. He watched from his window as huge machines entered the complex with their tanks filled with liquid gold.

As Jacob stepped off the car platform the long lines of cars reminded him of giant sand-dwelling worms basking under the translucent dome ceiling three hundred feet above his head. The electric motors of the idling Centas seemed to grunt in self-contentment.

Workers came along the tracks checking the tires for wear and proper inflation while others crawled under the huge cars, inspecting, poking and checking the apparatus.

Far overhead a crane slid on a gantry. A hose extended down from the crane while operators on top of the cars opened valves. Jacob stood and watched as a liquid was pumped from long lines of tank cars into an overhead holding tank.

It was obvious that the Kamsi had made a substantial investment in this system of growing and harvesting Solbean and transporting the processed liquid fuel to market.

"Doctor Ebbtide?"

The speaker was another well-fed Kamsi. Jacob wondered why all of them were so plump. Maybe they had learned to eat Solbean. "I am Sheik Gilla," he said. "This is my station. I am manager here."

Jacob nodded. Neither man made any effort to shake hands. Sheik Gilla kept his hands in his robes. Jacob kept his in his pockets.

A woman stood obediently behind Gilla. She was dark-haired and wore a veil. Jacob couldn't guess her age, because of her face veil and the thick black robe she wore.

"We must talk," Gilla said.

Jacob followed him out of the railcar and into a glassed-in office. The room interior was tiled in blue ceramics that reminded Jacob of the inside of a swimming pool. The air had a moist cold quality to it reinforcing the notion that the only thing missing from Gilla's office was six feet of water.

"Sit."

Gilla motioned Jacob into a chair and took his seat behind a large metal desk. Outside the entire operation of the terminal building was visible. Centas rolled into and out of the massive terminal. Cranes moved efficiently across vast overhead spaces pumping precious fuel out of mobile processing cars.

While Gilla was busy was some papers, searching for something, Jacob noticed the woman pouring water into glasses. She came by giving one to Jacob and the other to her boss. He looked up from the desk.

"I will be direct with you," Gilla said. "Omar is well respected. He deals with foreigners all the time. He usually gets results."

"I see," Jacob said, nodding and when Gilla said nothing more, Jacob continued, "But of course I agree. I was treated fairly by Omar."

That seemed to be what Gilla had been waiting for. He let out a deep breath. "Good now," he said, pushing papers across the desk at Jacob. "You signed a contract with the Kingdom."

"Yes," Jacob said, "I did. I signed a contract."

"Yes," Gilla repeated, pushing the paper over in front of Jacob, "but now we need you to sign an addendum. We need to modify the terms of your employment agreement." He handed a pen to Jacob who sat looking at the new contract.

"Why?" Jacob asked. "Isn't our contract still good?"

"As far as it goes yes," Gilla replied, "but you assaulted a nurse and your man was made to pay the price."

Jacob winced. He had hoped that all the unpleasantness about Hasham was behind him and that with new work and new people he could start over. He moistened his lips and thought that he heard the woman make a little laugh.

Gilla scowled at her. "Go to your desk!"

The woman said something in Arabic and left the office slamming the door behind her. When she had gone Jacob looked at his new contract.

It was the same contract he had signed before, the same piece of paper he had signed with Kasan, but now two additional pages were added.

"I don't understand," he said. "Why is this necessary?"

"Actually," Gilla said, "This is necessary because of our insurance company. If you do not finish your assignment they feel they should be compensated for their losses."

Again he made an expansive gesture to hand a pen over the desk to Jacob. Sheik Gilla held the pen up expectedly. When Jacob took the pen he sat thinking about his options.

He could understand forfeiting his pay if he didn't finish the work assignment, but compensate the insurance company for their loss? What were they in business for? This was 'sharp' dealing and he wondered momentarily if maybe the insurance company wasn't gouging him.

Still his contract clearly stated that he was entitled to one whipping boy, which as things turned out he had already used up.

"You mean," Jacob asked, "That if I don't finish my assignment I pay the insurance company two million gold dinars?"

"Plus interest and legal expenses. We are not barbarians," Sheik Gilla replied officiously.

"No, no," Jacob said, "Certainly not," but he didn't mention how Hasham's family had come to collect the insurance money in the first place. No, they were not barbarians only a modern day version of businessmen.

Jacob slid the papers around in front of him and signed. He reasoned that he was deep inside the Kamsi Kingdom and he had to make the best of it. Where could he go? Or, more likely would he be in any physical condition to go anywhere given the Kamsi Code of Justice if he messed up?

"Now," Gilla said, shuffling the papers and sliding them neatly into a folder, "That item is finished."

He stood up facing the glass wall that looked out to the terminal. Light and shadow mingled as the huge machines moved from one location to another. Technicians scrambled around on tank cars entering notations into laptop computers. Jacob stood up behind Gilla watching.

"You've done a good job," Jacob said. "You've reinvented the entire energy industry." For a long moment Gilla said nothing, but then without facing Jacob he pointed out at the terminal.

"We inventory and control all the movement of the Centas by satellite. We get field reports. We know what Solbean fields are ready for harvest. We enter that information into the system and it automatically assigns a Centas to make the run out to that field and process the harvested Solbean and deliver the fuel back here."

"Do the Centas report their locations to the satellite?"

"Yes. Where the Centas are on the ground and how much storage tank capacity they have left. The routing. Everything."

"Marvelous," Jacob said. "I am impressed."

He turned to see Gilla's expression darken.

"What?" Jacob asked. "You seem angry."

"I am not happy that, of all the people in the world, it should be you to come here to help us."

"Because of the accusations against my daughter?"

"We think she is involved."

"How?"

"We intercepted communications between the Americans and the Israelis," Gilla explained. "Her name and your name keep coming up. We cannot understand much of the conversation, because they knowing we are listening. They are, as you say, alert to the reality that their words are heard by us."

Jacob waited for Gilla to continue, but instead he looked out the window and then at Jacob.

"I need to know," Jacob said. "How do you come to believe that it's my daughter?"

"We know," Gilla said. "Besides Leviman works for us. He takes our money."

"Leviman?" Jacob asked. "Leviman?" he repeated. "Leviman works for you?"

Sheik Gilla nodded. Jacob's suspicions about Leviman were confirmed, but why would Leviman spread the word that Talya was involved even if she was involved? What game was he playing? Gilla bit his lower lip. Jacob watched his demeanor become angrier. Gilla shook his head.

"Leviman provides us with information, but it's never enough. Never enough!" he repeated angrily.

"Leviman and I are enemies," Jacob said. "And please remember," he added, "that I am the one who took control of the satellites away from the Americans in the first place. You couldn't grow any Solbean in confidence, because…"

"Yes, we know," Gilla interrupted him. "It's genetically engineered. The satellites were a problem. Yes, you solved that for us. And for that we thank you."

"The same satellites are still in use?" Jacob asked.

"The same?" Gilla asked, then, "I don't know, but the satellites come over the sky every night. The schedule does not change." He stood for a moment with his head down. "But we paid for the seed," he added.

"Are they killing the crop?"

"Not with the satellites," Gilla said, "but our whole system is broken. If not the Americans and Israelis then who?"

Maybe that was the American game, Jacob thought. Keep the twinkling satellites coming overhead, take some pictures, but do no *direct* damage. Not with the satellites. Maybe the Americans had decided it's better to attack in other ways.

Maybe the Americans had decided on another approach: Maybe sabotage the Kamsi control system and put the whole thing off on the Israelis? Talya, although young, was a known associate of Ben Leviman so that would be the Israeli connection: By attacking the Kamsi control system the Americans would work their damage on the crop and stick it to Israel at the same time.

"I believe my daughter is in danger," Jacob said. "I also believe she is used, because she is my daughter. By using her name you connect the damage to your crops with Israel."

"Yes. We think the Israelis are also involved," Gilla said.

Jacob turned to squarely face Gilla. "Let's put the past behind us. I think we can help one another. I really do."

"We know of your reputation," Gilla said. "That's one reason we have gone to such lengths to get you here. We are grateful to you. We have gone to great strides to get you here. We also believe that you can help us."

"I will help you," Jacob said. "I had trouble. I'm not the easiest man to get along with, but I am honest. I will help you."

"So be it," Gilla said.

"But to help you," Jacob added, "I must be given all the facts."

"Of course. It's that..." Gilla hesitated. He pointed out the window. "Getting facts is not easy."

"You have access to your own system?" Jacob asked.

"Yes," Gilla said, "but our control of the system has nearly broken down. We have Centas showing up in the middle of nowhere with no Solbean. We have Centas showing up here with nothing."

"Sheik Kasan told me," Jacob said. "The Centas wander off. The Solbean fuel, if any, is lost or stolen."

"Exactly." Gilla walked to his desk. "You see," he said, "We know when the fields have been harvested. We know the number of the Centas that processed the crop into fuel. The motion of the Centas is controlled by satellite. That's the problem. That's the problem," he repeated.

"Satellites are always the problem. If someone controls the satellites they control the system. And, if they control the system they control the Centas."

Jacob leaned back. So that was it. Somebody or some government was 'hijacking' the Centas, rerouting the trains to an undisclosed location, pumping out the fuel and sending the empty Centas into the terminal. Or into the Red Sea.

The Kamsi were the victims of massive well-organized theft. It had to be costing the Kamsi Kingdom millions of gold dinars a year.

"I will need to talk to your engineers," Jacob said. "We need to be very systematic and look at every detail. We need to trace back every ground signal sent to one of your satellites. I am confident I can help you get control of the system."

Gilla sipped his water. "There are no ground communications interfering with our satellites," he said. "No instructions transmitted from the ground."

"Nobody is sending signals to your satellites?"

"I didn't say that," Gilla corrected him. "I said no ground signals. Nothing. No signals, from the ground."

Jacob hunched his shoulders in confusion.

"There is constant communications between the Centas and the satellites," Gilla said. "We monitor all that traffic. All of the signals are, how do you say, *filtered?* Is that right, *filtered* through our system? We have checked. The Centas are sending accurate data to the satellites."

"And the satellites?"

"The satellites send accurate control information to the Centas."

Jacob stood up, walked to the middle of the office, thinking. "Is it always the same Centas that are hijacked?"

"No. We thought of that. All Centas are the same type of machine. Some can have passenger cars attached. Such as the one you traveled on.

"Sometimes the Centas can meet up in the desert and one can transfer fuel to another. We don't route all Centas into the terminal. Even some repairs are done at remote locations, but the control communications is located here. Satellite communication is all controlled from here."

"So repair crews at other locations don't have the control codes?"

"They have no access."

"And you are certain?"

"Yes, yes," Gilla said, "All signals from here are authorized. No illegal communication traffic with the satellites."

Jacob stood in silence thinking over the problem. The satellites had to be getting false instructions from a ground station. Or more likely somebody in a jeep was taking control of the satellites. Somebody had the satellite access codes.

Somebody, somewhere, *had* to be misdirecting the Centas and they *had* to misdirect the Centas by giving false instructions to the satellites and thence to the Centas.

"Sheik Gilla," Jacob tried to reason with him, "*Somebody* has to be sending false instructions right? There must be a breakdown between the Centas and the satellites. There *must* be."

"We have checked," Gilla said scowling. "We are not stupid! Our engineers have checked! We have aircraft in the air twenty-four hours a day!

"All communication traffic even telephone transmissions are monitored. There are no illegal transmissions to the satellites! No unauthorized signal traffic from the ground to space!"

Jacob grabbed the back of his chair staring at Sheik Gilla. "But there must be!" his said, his voice rising. "You're not…" He hesitated. No! No! It won't do to call his host '*stupid.*' No, no! Jacob cautioned himself. That wouldn't do, but any idiot knows that satellites don't make up their own minds!

Somebody had to be feeding new data into the satellite systems! Jacob looked around at the outer offices. People worked at computer stations while others milled around carrying papers. It looked like any office, anywhere.

"You changed the encryption code?"

"Of course."

"And the Centas are still hijacked? Hijacked," Jacob repeated, "The trains are stolen by a false instruction from a satellite?"

"We change the code, yes," Gilla said, "It matters not. The Centas are taken. They wander off. False instruction from the satellites causes them to drive off into the desert. Sometimes we find them with aerial photographs. Sometimes we don't. Sometimes the fuel is taken."

"Taken?"

"Spilled on the ground. The valves are opened. Sometimes the Centas drive in here to the terminal, but the fuel is gone.

"And you monitor the signals from the Centas?"

"Yes of course. Every transmission is recorded. We track each transmission down to one billionth of a second. We find nothing wrong. All Centas seem to be operating correctly."

"The Centas operate correctly and yet trains are hijacked?"

All Centas communications were encrypted and yet the trains were hijacked. Sometimes the fuel dumped on the ground and the empty trains routed back to the terminal as an insult to the Kamsi. The hijackers had a sense of humor.

But sometimes the hijackers felt like making more trouble so they parked the trains out in the desert. Somebody wanted the whole Kamsi system shut down and there was no doubt in Jacob's mind that it was the Americans probably working hand in glove with the Israelis, because who else had a motive?

The Chinese built the Centas and the satellites. Both Russia and China had agreements with the Kamsi; but besides all that, it was the Americans that had suffered from Kamsi terrorist attacks. If anyone wanted the Kamsi fuel processing system wrecked it would be the Americans.

America had vast desert stretches. After the beating they took with China and Free Trade it was unlikely they had the stomach for another bruising battle with the Kamsi over Solbean production.

America, the country that it had been, had fallen apart. The Americans were punch-drunk with wars and economic battles they couldn't win.

The Kamsi believed that they could produce Solbean and harvest and distribute the liquid product faster better and cheaper than the Americans. From what Jacob could see of their system they were right.

Now the Kamsi Solbean production system was in disarray. Everything pointed to the Americans and also to his own daughter. Jacob believed that the Americans did not need him nearly has much as he had thought. They were doing a great job ruining the Kamsi without any help from him. Jacob stared out at the terminal building feeling contrite and useless.

Once again he had over-rated himself. Once again the earth was spinning along without any help from him. He remembered being told once, by a professor at the University, 'There is no such thing as an indispensable man.'

He had spent the better part of two years slaving for Gladys Hines and not one American cop or engineer had bothered to visit him. No demands had been placed on Gladys to hand him over, but then why should they?

If the Americans were behind the hijacking of Kamsi trains they were doing a damn good job of it. He had handed control of two satellites to the Israelis who had handed control right back to the Americans. Ring around the rosy.

The Americans had, over the past year or so extended their control. The game for control had gone right on while he had been busy mowing grass.

Now, the Americans had taken the game to a new level, but how? How could they retake control of the satellites and therefore of the Centra without somebody detecting their activity? *Any* electronic signal can be detected. Jacob looked up as Gilla said something.

"Most Centas are robots. They can function in the desert for weeks without food or water or any of the things that people need. Most Centas have no humans on board. They are automatic machines."

"But," Jacob persisted, "The signal to the satellites must be coming from somebody. They must have access to your code key. The code is encrypted. It must be what we call an 'inside job.'

"Somebody in your organization must be selling the code or working for the Americans. It must be a passenger or a crew person that is riding on the Centas."

Jacob pointed to the busy office. "Or one of them."

"That is possible," Gilla agreed, "but the situation is more complicated. The signals received by the satellites are Centas signals, but it is not passengers or crew. As I said Centas that we use for harvesting the Solbean are completely automated.

"There is nobody on board. It is a machine. We have sensors to detect any living creature. We detect no living creature on the trains. Not even a field mouse goes undetected; and yet, the communications between train and satellites are corrupt. The system is broken."

Jacob understood that the electronic communications between the satellites and the Centas were two-way; it was a process, a 'handshake,' that made sure each partner in the communication process was bonafide.

So it wasn't just the satellites that were giving wrong instructions it was also the Centas that took the wrong instructions and acted upon them as if they were correct—

both the satellites and the Centas were subject to control from outside forces.

Gilla gestured to the office behind them. "We are almost certain that we do not have a spy in our system." He waited, his hands folded on the desk. Jacob knew Gilla was waiting for an answer.

The Centas were automated machines that, in some cases carry passengers and crew, but neither the passengers nor crew had any direct control over the machine. And most Centas that are in the fuel processing fleet have no people on board at all. As Gilla said not even a mouse rode the trains without detection.

The Centas' path was controlled from space, from the satellites; and, the signal controlling the Centas is encoded. Yet, the reality was that somebody was diverting Centas.

But if the signal wasn't coming from a ground station and if they couldn't detect any signal outside of their own activity then how in hell were the satellites sending false instructions; and, why were the Centas accepting those instructions and wandering out of control?

The Kamsi had had aircraft in the air twenty-four/seven. They monitored electronic traffic from around the world with no trace of bandit transmissions to or from their satellites.

"I know there's an answer," Jacob said.

"We will work together," Gilla said. "We will get you quarters," Gilla stood up.

"Then, later, I will introduce you to our staff. You have much work to do, but first it is time for the midday meal."

"Sounds good," Jacob said.

They walked out of the office through a network of administrative and engineering offices where people stopped their routines to look up as Gilla and the stranger went by. Once outside they walked to a play yard for the children.

Gilla pointed with pride to the old Space Shuttle, the NASA logo still emblazed on its side, with children rushing inside the huge cargo bay and sliding out, down chutes and slides, into a sandbox.

"We brought all the shuttles from the Americans," Gilla said. "We were lucky to get one of the Shuttles here for our children. Most are in Mecca and other scared cities."

He was obviously proud of his accomplishment. Jacob walked around watching the children and feeling Gilla's emotional satisfaction that he had the political clout to get one of the Space Shuttles for his own schoolyard.

In the lunchroom they ate rack of lamb, a salad, a delicious pastry and favored water. After lunch Jacob felt his stomach full. He followed Gilla out of the lunchroom to another building that was a tastefully appointed barracks.

Over the next few hours Jacob was given a new set of Kamsi style clothes and a credit card to be used for buying food. He sat in his room with a bed and a small window looking out over the terminal. He could watch the trains come and go. There was also a portable TV, but only one Arabic station.

He met with Gilla again and was introduced to some of the staff. This was his first introduction to working with 'rank and file' Kamsi citizens.

These were engineers who had gone to school in Europe, England, and the United Remnant. When Gilla was around they were polite and even smiled. When Gilla went back to his office, however, their demeanor toward Jacob changed.

In view of the dirty looks, contemptuous grunts directed at him, the reason for whipping boys was painfully obvious. Way down deep inside the Kamsi really didn't like western people and they didn't hesitate to show it.

'Tolerance' that Westerns praise so highly was completely absent. Jacob had the impression that these were healthy people that would not submit to the cultural browbeating so mercilessly applied to white Westerners.

By the end of his first day in dealing with Kamsi engineers it was obvious to Jacob that his situation was untenable. He would get no cooperation. The resentment was palpable.

Around him the engineers all spoke Arabic, although many of them also spoke French or English and could communicate with Jacob if they chose. The few sentences that Jacob did understand was that the Kamsi deeply resented this Jew, this foreigner, who was to come into their world and 'help' them with their problems.

They deeply and to Jacob's way of thinking justifiably resented this 'smart Jew' who had come to lead them out of their technical difficulties. That an outsider was to 'show them' the solution to how the Centas were being rerouted was an insufferable insult to their integrity and intelligence. They made no effort to conceal or sugar coat their contempt.

By the end of the first day Jacob knew that no matter how hard he tried, no matter how friendly he was, he would be accused; and once accused, all the efforts of Omar or Gilla, people that he now thought of as friends, would be no use.

At the end of the first day Jacob walked back to his room badly shaken. He sat on his bed stressed, not with physical fright, but with the knowledge that good engineering, good problem solving, required first and foremost the ability to concentrate on the problem at hand.

A good scientist needs to 'filter out' the rest of the world so that he can work on the problem to the exclusion of all else. Here, in the Kingdom, his attention must first and always be on the world around him. The technical problems would have to be a distant second. In the Kingdom he found himself in a crazy house.

Jacob laid down, half thinking, half asleep, trying to reason, fighting to find the courage to face his 'co-workers' the following day. He dreaded the sun peeking over the horizon in the morning. He had fallen asleep when a knock came to his door. Jacob got up and found Gilla standing there.

"Materials," Gilla said. "You must learn more about us. I brought you these materials."

Jacob accepted the books and English publications and thanked him for the materials on the Kamsi Code of Justice.

"For you," Gilla said helpfully. "This material will help you in dealing with our engineers."

Jacob took the materials. Kamsi had a tough and vigorously enforced Code of Conduct that only in the last week had Jacob learned to take seriously.

Thievery was punishable by the thief's hand cut off at the wrist. This punishment was carried out in public, often with others within easy splatter distance of the culprit, with the onlookers enjoying a basket of fruit. Every damn one of exclusively male audience seemed to be stuffing his face while watching the slaughter.

Jacob turned the pages wincing at the full color photos. After a few pages he slammed the book closed feeling moody again. Jacob knew that he would have to take every meal at the lunchroom across the schoolyard.

At breakfast, lunch and dinner, he would have to walk close to the children; and if one of the little darlings accused him of taking his lunch bucket or accused him of something worse there was a basket out there waiting for Jacob's severed hand.

Presumably, the good Kamsi would have lunch while enjoying the spectacle, the entertainment, he would provide as his hand or head plopped into a waiting basket.

Jacob picked up the materials leafing through the pages. Was Gilla trying to scare him? Well, if so he was doing a bang up job. Jacob stuffed the booklet under his bed and started to hyperventilate. Jacob had never suffered from claustrophobia or the feeling that the world was closing in on him, but now he struggled to control his gasping breath and his pulse. He found a copy of the contract.

He began reading every sentence as if his life depended on it. Only now, with a knowledge bred of fear, he knew his life *did* depend on it. He was still studying the contract when there was another tap at the door.

"May I enter," Gilla asked.

"Please."

He gestured the Sheik to sit on the bed as there was no other furniture. He and Gilla studied the contract papers that were spread out between them.

"I came back," Gilla said, "because I forgot to inquire of your first day. I saw you briefly before, but only to bring materials. Did you have a chance to study?"

"Very good," Jacob said swallowing as he did when he lied.

"You met staff? I am wondering how you are feeling?"

Jacob lifted the contract papers off the bed. "I do have some questions," he said, "Curious, you know, about the details.

"You know you advised me to read the contract. There's, ah, a few points I have trouble with it."

"What bothers you?"

"The part about the payback should an employee leave, that is, the part about the insurance pro-rate. The amount to be paid.

"It says, '…and to pay a pro-rata lump sum to damages done to other employees or to insurance carriers offering services and to shield from financial harm said insurance carriers such lump sum payment to be offered to Kingdom authorities upon resigning from its employ.'"

Gilla put his chin up with his eyes hard on Jacob.

"There's nothing difficult about it," Gilla said. "It is straight-forward English."

After an uncomfortable pause Gilla picked up the document. "The best way for me to interpret the contract for you, bearing in mind that I find it hard to believe that you don't really understand, is to put it in light of your situation."

"Please do," Jacob said smiling as nonchalantly as he could while staring back at Gilla and getting the unpleasant sensation that the Sheik was much better at fighting it out eyeball-to-eyeball than he was.

"In your case," Gilla explained, "Since another Kingdom employee Hasham lost his life and this fact has caused some financial strain that Kingdom authorities think it only fair that you would pay them back if you should choose to leave us."

"I get that idea, but ah, since we've using my specific case how much money would I be expected to pay?"

"If you serve out your contract time of one year absolutely nothing. The Kingdom is not un-reasonable, Doctor, but look at it from our point of view. If people came to us thinking that they could ignore our laws, our Faith, or way of life and make some hapless individual, the whipping boy, pay for their crimes and then expect to get up and leave without any obligation on their part why you can see there would be people coming here and trying to do that. You see?"

For the first time Jacob saw Gilla actually smile. A gold tooth sparkled brightly in the darkness behind his fleshy lips. While Gilla spoke Jacob tried to remember the web of events that had brought him to this dilemma.

Yes, he understood the bad feelings between Kamsi and other peoples. Kamsi grievances due to treatment they had received or thought they had received.

On the other side there was hostility of people against the Kingdom due to their riches and power and the attacks on other cultures and societies. There was hatred and grievance on both sides. Yet, here he was, stuck in the middle.

"I understand that I would have to pay money," Jacob said, "but how much? Look I want to succeed here, but what if things go wrong?"

"Well," Gilla said, "Hasham was with us for twelve years and that is good for you." Gilla took a pencil out of the folds of his robes and scribbled some numbers on the back of Jacob's contract.

"He was with us for twelve years and all whipping boys must retire at the end of twenty years. That is their employment contract with us." Gilla scribbled a few more numbers and seemed satisfied. He looked over at Jacob.

"Of course the odds of a whipping boy facing Allah is always the same no matter how many years of service. But Grand Sheik Rata did not know much about statistics. He thought that each year a whipping boy worked he was, as you might say, pushing his luck."

Gilla grinned, holding up the paper, but the room was too dark for Jacob to read. Jacob fumbled around in his robes to find his glasses, but came up empty handed.

"But now," Gilla said, "since Hasham had eight useful years left and since Lloyds of London will now have to pay off on a multi-million dollar contract it creates unfair results for the Kingdom, you see?"

Jacob knew he really didn't want to know. After his first day, with the attitude of his 'co-workers' Jacob knew he wanted out.

It only took him about five hours with Kamsi engineers to reach that conclusion. But, he needed a way out on better terms than paying huge sums of money over what now, to Jacob, had become an unfortunate misunderstanding.

As Jacob listened to Sheik Gilla he felt empty inside. The Kingdom was the last stop on his life's journey to nowhere. If he went to Israel he might be killed or jailed. He wasn't sure what the criminal charges would be, but Leviman was waiting for him; and, the United Remnant was an impossible situation.

Cut grass for Gladys for ten years, or dig a hole in soft dark earth to avoid radiation poisoning? In time become a trustee at Stonegate and take pleasure in terrorizing other poor bastards entering that radioactive wasteland for the first time? Jacob looked back as Gilla studied his numbers.

"How do things add up?" he asked gulping. "If Hasham had eight years left how much money does it mean I should be responsible to pay?"

"Hasham insurance premiums were one hundred thousand gold dinars a year," Gilla said, "and if that seems a trifle high…"

"Trifle?" Jacob asked.

"We are obligated by our contract with Lloyds to continue to pay the insurance premiums even though Hasham is no longer with us."

"You pay insurance premiums on a dead man?"

Jacob's head spun with the financial implications. No wonder Gilla was still scribbling numbers on the contract.

"Lloyds of London was the only Company that would offer insurance."

Gilla turned the paper around to Jacob, but the light was dim. All Jacob could see was some chicken scratches in Arabic and a series of neatly printed numbers that seemed to have way too many decimal places.

"They gouged you," Jacob said to which Gilla smiled broadly, a finger coming up to his head, tapping his temple.

"Yes, we know. Remember, Doctor, the insurance company has a very high payoff. Also, you must remember that Lloyds is a British firm and, of course, they keep their rates high, as you say, to gouge us."

Jacob thought that the British may or may not have thought too kindly of the re-emergence of whipping boys in the world—some two hundred years after they discarded the practice—but there was no doubt they would play the insurance game any way they could.

The British needed to make up for the billions going back into Kamsi hands for oil and Solgas. The English and the Kamsi were playing hardball and Jacob felt as if his butt was first base with both teams running to home plate over his backside.

13

"So it comes down to eight hundred thousand gold dinars to compensate the Kingdom plus two million to Lloyds of London."

"And of course the medical care for the harm you did."

"Any number of that?" Jacob asked with his voice quivering.

"About a hundred thousand gold dinars," Gilla said still smiling, "Expensive head."

Jacob could read Gilla's attitude clearly. The Kingdom's position was perfectly reasonable. So long as people came into its employ they were given protection in the form of whipping boys. They had protection from a cultural Code of Justice completely foreign to them. So wasn't it reasonable to balance the proposition out?

Shouldn't a European that takes advantage of a situation be expected to make some form of compensation? Furthermore, the Kingdom reasonably waived all such payments if the terms of contract were done.

If the employee put in one year's service all was forgiven. The Kamsi met all obligations to the insurance carrier.

The employees pocketed their paychecks and left. The Kingdom insisted on payment only on the condition that the employee succeeded in getting his whipping boy killed and then decided that, since the fun was over, to leave before doing anything useful to make up for the fuss he'd caused.

"Isn't that fair?" Gilla asked.

"I suppose," Jacob said suppressing a groan. "It is fair."

"Then, there is only one thing more that an employee might want to consider," Gilla said. He put the contract papers down on the bed.

Jacob had the impression now that Gilla was figuratively shoving him out of an airlock into deep space with no spacesuit. In response to Jacob's nod Gilla added,

"The consequences of quitting without having the money to repay."

"Credit's no good huh?"

The Sheik laughed out loud. "We do cash business."

Gilla's laugh turned mean. Jacob was now sure that under his flabby exterior was another little man in a white butcher's apron.

Gilla reached over to the materials he'd left. He flipped through the pages, twisted the book back on its spine pointing. "Here," he said. "This is the Code."

Jacob stood, thanked him and after Gilla left his room Jacob sank back to his bed. He wondered what would happen if he tried to run and failed. Jacob wished for the homey comforts of Stonegate. A hot meal, a cold shower and a hole.

Maybe living in a dark hole wouldn't be so bad after all. Animals had lived in holes for ages. The whole damn human race was heading into holes. Nice, comfy warm holes.

Jacob put his face down in his hands trembling with the thought that he could be a pioneer along with the other outcasts at Stonegate. A pioneer leading humanity back into the earth. His hands trembled as he picked up the material Gilla had left. Flipping open the booklet, he read the material that Gilla had left for him to read.

"Yeah," Jacob said sadly. "Some surprise."

The penalty for running was a long term of imprisonment. Jacob flipped through the pages looking at the conditions in Kamsi jails. They made the loft at the Hines Prison House look like paradise. Among others things there was no apparent air conditioning. My god, Jacob mused, "They may as well burn me alive," and then he flipped the page.

In the UR any infraction of prison rules brought more time or loss of privileges, but in Kamsi prisons any infraction of the rules brought torture including branding. Jacob sat on the bunk hyperventilating. Running was not a good idea.

He tossed Gilla's educational materials to the floor. He lay on the bed thinking. Somewhere along the way he had picked up the idea, he didn't know where or when, that anybody who destroys something and then willfully tries to avoid consequences of it is no better than a thief. He had lost his temper and had, for all practical effect, killed Hasham. For the second time in forty years Jacob prayed.

"God. If you're there, can you hear me? I'll do this work if you let me, but you're going to have to make some changes. Run some interference for me will ya? You know I'm not here for the money. I want to make things right. I understand plant genetics and because of the age we live in I have had to learn other disciplines.

Studying to keep up with my own field of study I also had to learn satellite communications. And then, god damn it, encryption!"

When Jacob realized he was cussing in his prayer he stopped praying. It seemed as if he had spent his life educating himself. He had studied and strived.

He had worked. And where had it all gotten him? Here. A hireling in the Kamsi Kingdom! Jacob closed his eyes. He was lying in bed fully clothed. He had learned many years before that a good night's sleep often helped to clear the mind and to solve problems.

He stared at the darkness. A gray twilight crept through the translucent dome of the terminal building. Thin gray shafts of light filtered like a thin veil over his bed.

A strange quiet had settled over the terminal during the night. The wind died down. Jacob listened to the rhythms of the night and sensing the world around him.

He felt a strange pulling at his heart, but knew he wasn't having a heart attack. In spite of his fear he felt a strange affiliation for the Kamsi, for this place, even for the desert.

When he saw the sun's morning light, a subtle coloring from gray to yellow, diffusing through the terminal dome Jacob laid in bed thinking. He made a resolution.

He felt a change of heart. Yes his situation was grim, but not impossible. He resolved to make the best of it. As yellow diffuse sunlight slowly worked into his room he felt his spirits rise with the sun. For the first time in memory Jacob felt optimistic and even enthusiastic about his prospects.

He would face his situation squarely as a man. He was through with running, with dodging, with playing games! He felt gratitude to the Kamsi for teaching him to act like a man.

In America, if he'd knocked out a nurse and attacked a doctor two sets of lawyers would retire off the money he would have to pay them. Justice would be slow in coming. He would pass time as he rotted in a jailhouse probably mowing grass for Judge Hines.

On top of that everybody else would be a tax slave made to pay for his upkeep. Then, on his release, Gladys would give him a bill, due and payable, for his incarceration.

He realized that there was something to be said for the Kamsi way of doing things. All he had to do was stay out of trouble for a year. For a year. For a year.

He put his feet on the cold stone floor and smiled with a renewed sense of purpose. Yes sir and yes ma'am. He would see it through. Right through to…a knock came to his door.

Jacob suddenly moaned, rolled over onto the bed and buried his face in his robe as a sudden pang of doubt seized him. The enthusiasm he felt only an instant before was gone, instantly driven away by a simple tap on the door. Getting up, opening the door, Jacob saw the Sheik standing there.

"We must see you immediately," Gilla said.

"Now?"

"Yes."

Jacob got dressed and drew a dark silk sash around his waist. His robes fitted close to his body he sat on the bed and slipped his feet into his sandals. He was even getting used to the free-flowing robes. He followed Gilla out of the barracks across the schoolyard to the terminal office building.

"In here," Gilla said opening a door.

The interior of the room was large and dark. Jacob entered the conference room and let his eyes adjust to the dim light. A large oval table dominated the middle of the room.

At the far end from the door sat two men. He didn't recognize either of them, but they had on western clothes not robes.

Sheik Gilla led Jacob down the length of the table and gestured him into a chair near one of the waiting men. Both men rose politely.

"I'm Hancock," one of them said.

He was about Jacob's age gray-haired and very thin. His handshake was firm and Jacob wondered briefly if maybe he had long experience pushing lawn mowers. His face was intelligent, but his eyes were hard. He could be a scientist, a cop or a con. Jacob looked to the other man.

Sweat soaked through his open shirt. He was beefy and completely unsuited to life in hot climates. His fat kept the heat inside his body and it showed. His shirt was ruined.

"I'm Suboy," he said. "Hancock and I are bounty hunters," he paused for breath and added, "but we do other duty like prisoner retrieval."

"Prisoner retrieval?" Jacob asked.

When Gilla sat at the other side of the table the other man reached over and at first Jacob thought he intended to shake hands. Jacob reached across the table with his right hand, but let it hang in the air as the second man clicked handcuffs tight around Jacob's wrist. For a moment Jacob sat with his hand reached out across the table surface, his mouth open and his mind blank.

"What is this?" Jacob asked.

"Americans," Gilla said. "They have paid your obligations," and before Jacob could respond the Sheik added, "Isn't that what you wanted?"

"I came here in good faith!" Jacob shouted.

He pulled hard on the handcuff only to find that the other cuff was on the fat American's wrist. Mr. Suboy was far too heavy to drag across the table.

"Why?" Jacob asked. "What happened?"

Gilla slowly got up. "I spoke with my engineers," he said. "They will not work with you."

"Now listen!" Jacob shouted, "You knew when you hired me! You knew I wouldn't be welcome here. That's why I had a whipping boy in the first place!"

"Yes, but now you don't."

Gilla stepped away from the chair. "With Hasham you would have a chance here, but without that protection, I believe your situation is impossible. I'm sorry, but I believe they will make trouble for you."

Jacob lowered his eyes, because that was the exact same conclusion he had reach last evening. He thought of the schoolyard with the Space Shuttle converted into a slide for laughing Kamsi kids and severed hands plopping into baskets. Jacob thought of the remarks and the looks he'd gotten from the Kamsi engineering staff.

"Judge Hines wants you back," Gilla said. "She found the money in her budget to reclaim you. You're going back."

Jacob was close to tears, "but how did she..."

Gilla cleared his throat. "The American Government contacted Judge Hines about a report they received when you escaped from prison," Sheik Omar said. "That was over three months ago. They found out from a man named Kelly that Kasan was sending you to the Kingdom."

"You didn't give me a chance!" Jacob shouted.

"At least Omar gave me ninety days. He gave me a chance!"

"Yes and it cost the Kingdom a good man! Give you a chance?" Gilla repeated. "You got Hasham killed! You asked me last night how much money you would pay if you left. I asked our engineers if they could work with you. They said no. I think, Doctor, they dislike you even more than most foreigners." He motioned to the bounty hunters.

"They have been here for over a month! Waiting. When I spoke with our engineers last night I knew. Yes I knew!"

Jacob blinked. His watery eyes went to the two men. Hancock slowly stood up gesturing Suboy with him.

"You think you're the first engineer that couldn't cut the mustard over here? You're the fifth or sixth one we pulled out of here this year. We spend six months a year flying you bums over to Prussia."

"I'm not a bum!" Jacob shouted. "How dare you do this?" Gilla looked away obviously embarrassed by Jacob's display of feminine emotion.

"No you're not a bum," Suboy said. "You're special. Because you're so damn special we have to take you all the way back to the United Remnant! We have to baby sit you for ten miserable days!"

Jacob ignored his outburst by forcing his disciplined mind to understand his situation. Jacob straightened up. When he looked over at Gilla he thought of the numbers written on his contract. So that was it!

"I understand," Jacob said.

It was only a matter of time before Gilla turned him over to the Americans and pocketed the money. Apparently the Americans had an extradition agreement with the Kamsi. It was only a matter of time.

All scientists were contract employees. Not citizens of the Kingdom. Once their useful work was done the Kamsi had no further use of them the Kamsi might pay the contract or might decide it was better business to frame the engineer for a crime and let the Americans pay the freight back to the United Remnant.

Even if he did the full tour, finished all the work, when the year was up, he would leave in handcuffs. The Kamsi got the work done and then made more profit by selling him to the Americans. Westerners were cash on the hoof.

"You're in bed with the devil!" Jacob shouted.

"You've made a pack with the devil!"

"The Americans are still our biggest market," Gilla explained, "even if we compete on Solbean. Bus…"

"No!" Jacob shouted as Suboy pulled him away from the table, "Don't say that! Don't say that 'Business is business!' Damn you!" Jacob cussed, "Don't *you* say that!"

Gilla shrugged. "But of course it is. As the Americans say, 'The world goes on.' Right? It is merely business," Gilla said. "This is a matter of contract. A Jew should appreciate that!"

Jacob staggered away from the table nearly tripped over the chair as Suboy showed surprising speed for his girth. Hancock took Jacob's other arm.

"Come on!" Hancock said leading the way out of the conference room. Jacob walked between the bounty hunters. Suboy threw open the door and a blast of hot air rushed in enveloping them. Already in the early morning the heat was insufferable.

Outside the building the sun hit Jacob's eyes. They walked around the side of the administration building with Sheik Gilla following. Jacob was taken to a waiting vehicle.

"It was not to be Doctor Ebbtide," Gilla said. "You could not help us. You have knowledge. Of that I am sure, but that is not enough."

Jacob heard kids laughing at the Shuttle slides and the wind coming into the compound off the endless desert. Then he heard the vehicle's engine start. Once inside handcuffed next to Suboy the sedan lurched away toward the terminal where, Jacob was sure, a Centas waited that booked passengers.

The yacht's sleek hull and teak decks gleamed white against a slate gray sea. Doctor Veeby admired the Sheik's yacht from his car window as his limo came to a stop. The driver, a long time trustee dong life for spying for the Russians, got out and opened the rear door.

"Wait Hansen."

Warden Veeby walked to the edge of the dock. Below him a two-man crew waited with a motor launch. They let the engine idle.

The put-put engine noises made Veeby almost weepy with nostalgia. There was a time not so many years ago when people routinely let small engines, even automotive engines, idle.

It was evening in autumn. A beautiful Indian summer day in what used to be New England. The air was fresh off the Atlantic. Whitecaps surged under the dock, rocking the motor launch with dark rushing water slurring the engine noises.

Veeby walked to a wooden ladder, tested it, and swung himself over the water with his shoes awkwardly finding each rung.

Finally, after struggling down and enduring dirty looks from the crew, he stepped onboard the launch. Five minutes later he was on the yacht's deck. A man and a woman were waiting at the railing. The man had, like the crew, his head wrapped in a headband.

"Another rich Kamsi," Veeby muttered under his breath.

As Veeby pulled himself up he looked hard at the woman. She was slim, in her fifties, mostly gray haired and was used to command. She stood with the demeanor of one accustomed to giving orders. When Veeby reached the deck she turned abruptly and walked away. The man wearing the turban came forward.

"Doctor?"

"I'm Doctor Veeby."

"I'm Sheik Halim Kasan. I represent Sheik Gilla here in the UR."

Within moments Veeby and the Sheik were comfortable in the lounge. A servant brought drinks and a large bowl of fresh fruit. After the formalities and general comments on the weather Veeby decided it was time to get down to business.

"Why did you invite me?" he asked.

Veeby sipped his drink while taking in the well-appointed layout of the lounge. The gently rocking ship relaxed him. He took a calm delight in watching a lovely day slipping into dusk.

"Not that I don't appreciate it," he added, "But, you know, curious."

The Sheik didn't answer immediately, but seem to consider some private matter. After a moment he sat his drink on a table and put his hands on the armrests.

"We are the people that paid to get Doctor Ebbtide out of your prison," he said in a matter of fact voice.

Veeby studied him unsure of where this discussion was going, but Veeby realized at that moment that his invitation was not a social call. The Kamsi were up to something, but what? Veeby decided to play along.

"I see." Veeby crossed his legs defensively.

The Sheik moved closer to the seat edge. His face conveyed anxiety. His tone of voice also belied his nervousness.

Veeby pegged the Sheik as a sensible man. As warden of Stonegate he could have this man arrested before he got his boat out of New Boston harbor.

"We paid to get him out," Kasan said again.

"Facilitating a prison break is a crime," Veeby said calmly. "Perhaps you should reconsider what you are saying. Helping to break somebody out of jail is a serious crime. I can have you arrested."

"But I don't think you will," the Sheik replied.

The Sheik seemed to gain a measure of self-confidence. Veeby put his head back in sudden understanding.

He had merely *threatened* to have the Sheik arrested. That was a lame bluff and the Sheik knew it. Had he been serious Veeby would have picked up his cell phone and called the New Boston police. Veeby resolved to play a better game. He watched Kasan carefully.

Now, when the Sheik took his drink his hand didn't shake. "Perhaps," the Sheik said, "Yes, you can have me arrested, but possibly you can answer a question for me?"

"I doubt it," Veeby replied, but still wondered what game the Sheik was playing, "but you can ask."

"It's simple."

The Sheik finished his drink and then gave a short instruction to a servant who left the lounge. Then, with his attention on Veeby he asked, "How did Doctor Ebbtide get out?"

Veeby stood up. His visit was over. There was nothing more to talk about. "You can petition the Bureau of Prisons for a copy of the report."

He stepped away from the chair, heading for the lounge doorway, but then the gray-haired woman that he had seen on deck entered. She stood silently by the doorway watching him. Sheik Kasan rose respectfully, acknowledging her, but spoke only to Veeby.

"We need to clear up this situation."

Veeby put his hand on the latch. "You've admitted to a crime. You should leave the UR."

The Sheik seemed intent on his own accusations.

"We have evidence that you assisted. We paid money. You took our money. In our society that makes you our employee!"

Once again Veeby detected uncertainty. The Sheik's voice broke even as he was trying to threaten. The Sheik, Veeby decided, was an amateur.

"We are not in the Kingdom," Veeby replied confidently.

He opened the door leading out to the deck, but two heavily armed men barred his way. Veeby stood staring from them to the deck and to the water.

He knew he had made another mistake by attempting to leave. Now, faced with these two men, he was again playing the Sheik's game. Men like the Sheik were brave so long as they had muscle to back them.

The yacht motored slowly away from the dock and was already out in the harbor. Silhouettes of dark buildings receded even as the sun sank blood red behind them. Veeby slammed the door shut, turned around to confront the Sheik, while pulling a cell phone out of his jacket.

"Turn this tub around!" Veeby commanded.

"I'll have the Coast Guard out here in ten minutes!"

"Doubtful," Kasan said. He folded his arms and waited for Veeby to reply.

"What stunt are you pulling?"

Kasan poured another drink into his glass, completely ignoring Veeby's remark and equally ignoring the woman at the other end of the lounge. She stared at Veeby, but said nothing.

Veeby slowly walked back into the lounge. His gaze went from the woman to the Sheik, who handed a drink to him. Veeby reluctantly accepted it. When the Sheik looked pointedly at the seat Veeby sat down.

His mind wandered over the many possibilities that included that he might be a kidnapping victim. While kidnapping was big business in the UR he doubted this Sheik would be so directly involved in a petty crime. People with money hire dirty work done. A man like the Sheik wouldn't soil his hands with it. So, if it wasn't a kidnapping, what?

"Let me explain," the Sheik said.

When Veeby looked again at the woman standing by the door the Sheik shrugged his shoulders. "At this moment, Sir," the Sheik said, "This is between us."

"There's nothing between us."

"Oh yes I think there is," the Sheik said, but before Veeby could make comment, he added, "You know Doctor Jacob Ebbtide?"

"I remember him," Veeby said. "He escaped."

"You had nothing to do with it?"

"Of course not."

He looked out through large tinted windows to see that the yacht was moving majestically through New Boston harbor on its way out passed a massive stone breakwater into the Atlantic.

"People know I'm on board," Veeby said. "My driver will report that I am missing."

The Sheik smiled expansively. "But you're not missing are you? You are right here."

"Turn this tub around!" Veeby said again.

Slamming his drink down to the ornate table surface he started to get up again, but as Kasan sat back, watching him, Veeby thought again of the two-armed guards. He calmed down. This Sheik won't do anything, Veeby reasoned, even if he did have two thugs outside.

"Doctor Ebbtide left the Kingdom eight days ago. He was in the Kingdom a little over three months, but he had trouble. He caused the death of a man. It was a mistake to hire him. He has cost us much money. Now he is coming to the UR.

"In two or three days he will be returned to the UR. He is being flown back at our expense in the company of two men."

The Sheik looked at the woman. "We are reimbursing the Americans," Kasan said. "Travel these days is quite expensive, but we have the money to help the Americans."

Veeby was unmoved by this information. So? An escapee was coming back. Nobody would take Ebbtide's word for anything. After all he was an escaped felon. Besides, Ebbtide would have to be crazy to open his mouth about how he escaped. Veeby let himself make eye contact with Kasan.

"So?" Kasan asked. "You admit to harboring him?"

Veeby tried to gauge the Sheik's reaction, but he could detect no hint that violating UR law meant anything to him. Veeby understood that violating UR law was a national pastime. The 'Law' was a laughing stock.

Announcing one's self a lawyer was an invitation to be mugged. Stupid was the attorney that actually hung out a shingle. Veeby thought about Ebbtide.

As a returned felon he would be returned to prison and very possibly to Stonegate and right into Veeby's waiting arms. Veeby let his imagination wander. He felt satisfied that he could control any damage Ebbtide or this Sheik might do.

"I want you to see our problem," the Sheik explained. "Doctor Ebbtide is a unique man. He has unique knowledge. He can do us a lot of good. Or a lot of harm."

"At the risk of stating the obvious," Veeby asked, "Why not kill him?"

The Sheik frowned. "Because we need him," he said.

"It's that simple. Our plan was to bring him to the Kingdom. He was to work with us to solve some highly technical problems. But he is a difficult man. It didn't work out. The nature of our problem does not concern you."

Veeby looked out the window. The yacht was motoring out beyond the breakwater and already he could feel long rolling swells of the Atlantic under the keel. Veeby put his legs out, twirling his drink glass between his hands. His mind turned to other possibilities. It made him feel comfortable.

If the Sheik wanted to harm him the guards would already be inside the lounge. So, the Sheik wanted to talk, to bluster and to bluff. No rough stuff. That was fine with Veeby.

"So you didn't kill him, because you have another plan. You need his expertise to help you solve your problem. Right?"

"Yes," the Sheik said. "And we need your help."

Veeby leaned forward, lifting his empty glass and politely acknowledged the Sheik as he refilled it.

"And the plan?"

"It's very simple. You already work for us. Doctor Ebbtide is coming back here. He knows you. He trusts you, because you helped him escape." The Sheik voice hardened.

"Yes! We know! Don't deny it!"

Veeby met this challenge head on. He sat his glass down on the table hard. As his drink spilled he raised a hand to the Sheik.

"What you think you know and what you can prove are two different things! Don't forget it!" He raised a threatening finger to the woman. "And that goes for you!"

The woman angrily folded her arms over her chest. She stood by the door and said nothing. As the Sheik spoke again his voice carried the same weight and same tone as if he were speaking to a servant.

"Ebbtide trusts you! You will tell him that the Government wants his help. He will think he's working for a special department, but that he must always be careful, because he is still wanted by the law. His handlers will claim that they can't fix his legal problems, because of his wife the Judge.

Understand? He will think he's working for them when, in fact, he will be still working for us! That's how it must be!"

"What?" Veeby asked looking at the woman with a new understanding. "His wife?"

"We're divorced," the woman said. She scowled at Veeby. "I suggest you listen to what this man has to say."

"We need Doctor Ebbtide," Kasan said again.

"And when his work is finished?" Veeby asked.

The Sheik looked over at the woman. "Doctor Ebbtide has broken the law. He has obligations. There is nothing we can do about that."

Veeby didn't know Ebbtide at all, but had done some checking. While Ebbtide had a 'checkered' past Veeby had found out about Ebbtide's connections with the O'Hannon outfit. Even so he was clean compared to these damn Kamsi.

Veeby decided that this Kamsi best not go for a walk in New Boston. He drew the line at having Kamsi come into his country and turn everything to their advantage.

As Warden of Stonegate he could send this Sheik to prison on any number of different charges. And, he'd make damn sure this Kamsi deeply regretted threatening him. Before he could talk the Sheik again waved a finger to silence him.

"Listen now," the Sheik said. "I'm telling you something. Listen!"

"Make it quick," Veeby answered forcing to keep his voice down, "I'm late for dinner."

"We have made arrangements with the UR government." The Sheik turned indicating the woman. "Ebbtide will be put back in prison, but it will be a special prison. A private prison. We have made arrangements!

"He will be working for us, but he will think he is working for the Americans and the Israelis. And you," Kasan pointed at Veeby, "will be there when he comes into the UR to bring him to us!"

"You don't need me!" Veeby insisted. "Arrest him! Incarcerate him!"

"He trusts you," the Sheik insisted, "and besides he has something on you. He can make trouble. You helped him escape! Don't deny! Ebbtide will believe that you are helping him to escape again. That you are helping him to cover your own criminality! He trusts you. He won't believe that others are helping him. He thinks that you will lead him to safety! That's all you have to do!"

Veeby nodded. Now he knew the game. He was a key piece in a Kamsi scheme to get work out of Ebbtide. All he had to do was lure Ebbtide to their location. Other Kamsi operatives would convince Ebbtide that he was working for friendly Government scientists. They would 'play' to his ego.

They would tell Ebbtide that he was too important to end up in prison. Mistakes had been made, but now they were correcting those errors. They would tell the Doctor that he couldn't risk recapture. Ebbtide would sink into a soft captivity.

There would be no bars on the doors. No ham-fisted guards to beat him. He would have 'friends' who would clear certain activities for him. There would be safe restaurants.

Maybe even a safe park where he could feed the birds. And yet, Ebbtide would be surrounded, controlled, almost as certainly as if he were back in Stonegate. Ebbtide had information or expertise that these Kamsi cruds needed.

The Kamsi needed him, but found him difficult. Something had gone wrong, but they still had the idea that they could use him. Hmm. Veeby let his thinking wander through the possibilities. There was room to maneuver here, although this Kamsi seemed to think he was in a corner. He came to a conclusion. Yes, now, that was it. Veeby knew he had to play a hard game.

"Dream on!" Veeby shouted jumping to his feet. "I'll have him arrested if he sets foot in the country. Ebbtide will be right back in Stonegate!"

"You have his confidence!" the Sheik shouted while also getting to his feet. "We have much time and money in this! You will lead him to us. Then we can control him!"

"How many different ways can I say it?" Veeby shouted, coming nose to nose with the Sheik, "Go to hell!"

Kasan shouted something in Arabic and the door behind the woman opened. A man entered, but in the evening's failing light Veeby couldn't make out his features.

The man stood to one side of the woman, slightly behind her, facing her with his hands down at his sides submissively. It was obvious from his behavior that the man knew the woman. She looked at him with a gaze that would melt steel.

"What have you to say for yourself Kelly?" she asked. Veeby's eyes opened wide in recognition as he watched Kelly moved cautiously, respectfully, around the Judge. He came out of shadow into the light.

"The Warden helped us escape Judge. I'll swear to it in court. He walked us out. Ebbtide knew the Sheik and got the money for our break."

'Judge,' is that what he called that woman? Veeby's heart rate ratcheted up another notch. Before Stonegate Ebbtide had been housed at the Hines Prison House. She had a reputation for being tough.

"The gold coins were marked," the Sheik said.

"We traced those funds to your bank account. We can show that you took the gold and we have this man, Kelly, who will swear it was you that helped them escape."

Veeby sat down, but now was more puzzled than worried. The woman called Judge came up to the Sheik.

"Satisfied?" she asked. "I had nothing to do with my ex-husband breaking jail!"

"Yes," Kasan said. "We know the situation. Still, now with this new event. We had to contact you, because we knew Mister Kelly was now returned to your custody."

"Keep me informed," she instructed him. Veeby noticed how submissive the Sheik became in her presence. She had the knack of scaring the hell out of men. She had a flat-footed way of standing that made Veeby want to crawl to her.

"You will be completely informed," the Sheik said.

"When he has finished the work for us he is yours."

"I'm satisfied," she said. "I won't interfere. Make sure you keep your bargain." She motioned to Kelly.

"He is my eyes and ears."

Veeby remembered the name Kelly from the papers. He had killed the last Mayor of New Boston and his wife; and, he was serving time, but not at Stonegate. Why? As Veeby wondered about Kelly's luck in getting preferred treatment in the legal system the Sheik yelled something in Arabic.

A servant entered, stopped by the door, holding it open for the Judge. She walked slowly to the doorway and started out, but hesitated.

As she door swung open fresh sea air breezed through the lounge. The salty mist, the brisk air, enveloped Veeby. He breathed deeply, felt alive, and was glad that things had turned interesting. The woman stood in the open doorway. "Did you get them out?" she asked.

"No." Veeby lied and stood facing her. "I won't screw up a good job. Not in this economy for a few dinars gold or otherwise."

The yacht rolled slightly in a gathering sea and as the ship lolled to one side Veeby grabbed the table. He watched as Kelly rushed up and took the Judge gently by an arm, but when she jerked her arm away Kelly retreated almost instantly. There, Veeby knew, was one damned worried man. That poor pussy-whipped slob was completely at the mercy of that woman and he knew it.

"But Stonegate was considered break-proof," she said.

"The only way out is by air or by lead box. It's been written up in all the papers."

"We don't check if anybody is gone," Veeby said.

"Hell, who cares? It's nothing but radioactive waste out there. If they want to venture it fine. Who cares? Ebbtide is smart. I guess he made it." He raised his hands. "But who really knows? Maybe he'll be dead from radiation in six months."

"You took us out Warden."

Kelly didn't venture out from behind the Judge, but he spoke confidently. "I'm not sick from radiation. You left a way out for yourselves."

"Sensible," the Judge said. "Is that how you did it? Walked them down the garden path? The only safe way out of the crap?"

Kelly came around the Judge, but stood to one side as to not block her line of sight. When the Judge shifted her weight from one foot to the other Kelly cringed and moved further away. Kelly stood behind her like a slave awaiting orders. Veeby brought his face up to directly face his accusers.

"Your word, a confessed killer, against mine?"

Veeby felt confident he could out bluff these Kamsi and their hangers-on. After all, it wasn't as if he didn't have big political backup. He was a national hero. The man that discovered the public was being poisoned by Kamsi oil. Let them threaten! He could always claim political retribution. When push came to shove this Sheik would be swimming for home. Besides, as his old man always told him, *never confess*.

"And Kelly," Veeby added, "Don't ever find yourself back in Stonegate."

"Kelly doesn't have to worry about that," the Judge said.

"And what about you Kelly?" Veeby asked. "So you've done some checking? Good, because so have I..."

He looked to the lounge windows, timed his remarks and let his sight fasten on the Sheik

"I'm glad you invited me out here," Veeby continued, "because I've done some checking of my own."

He pointed to Kelly. "Has anyone asked how a guy that murdered his wife and the ex-Mayor ever got into a sweet little boutique jail? Huh? Was it luck?"

Kelly stepped slightly behind the Judge as if seeking protection of a severe Atlantic storm. The Judge stood her ground, but looked around aggressively. That was a good sign. She probably knew good and damn well that there was something wrong with a legal system assigning a hardened criminal like Kelly into her homey little prison.

"All violent felons are my guests at Stonegate. Yet, somehow you got a suite at a house party. How come?"

When Kelly stood with his gaze downward not venturing to look at Veeby the Warden looked at the Judge.

"He was your guest? Right?" Veeby asked. "Am I right about that? That mass murderer got a billet at your place. Right?"

Judge Hines seemed genuinely surprised. She spun around taking Kelly's shirt in both hands tugging on him.

"Tell me!" she shouted.

Kelly stood dog faced as Kasan spoke.

"Yes. He played a part in our arrangements. But you must understand Judge Hines we knew that you had been married to Doctor Ebbtide. We made sure the local police knew his arrival date and time."

"You arranged for him to fly into New Boston? Where our local police would capture him?"

"Most certainly. We routed his flight through Prussia. We made sure his flight had fuel. We arranged it. We spent a lot of money to insure that he found himself in your jurisdiction and in your court. We felt certain that, given your problems with him, he would be a guest at your facility.

"We had a complete psychological profile done on you Judge Hines. We know you quite well. You see," the Sheik nodded at her, "We were right."

Veeby watched the Judge's face tighten. Her eyes squinted first at the Sheik and then at Kelly.

"You arranged for Jacob to fly into New Boston?"

"Certainly."

Judge Hines gestured to Kelly. "And him?"

"Certainly."

"Why?"

"Because, we were certain that you would make Ebbtide's life miserable. We did a profile on you. Remember? I just said so. We felt certain that you would punish Ebbtide to a point where he would have to seek us out. You see?"

When the Judge grunted in astonishment the Sheik added, "Sure we could have kidnapped him, but he would not cooperate. We need his willing assistance. We needed him to come to us! To come to us," he repeated. "Understand?"

"You used our legal system to get at Jacob?"

The Sheik raised his nose in the air. He assumed an air of self-satisfaction. "Well," he said defying her to argue the point, "Our plan worked to perfection. Didn't it? It didn't take him long to run into our arms did it?"

"What about him?" she indicated Kelly.

"We contacted Mr. Kelly after his arrest. We promised him easy jail time if he would look after our man. We are not fools Judge Hines. We needed to protect our investment. We needed someone like Kelly. He is tough and desperate. You see it's critical for us to keep Doctor Ebbtide alive and healthy. He is our investment! You have no idea how valuable he is to us. We knew, sooner or later, he would contact us. And, when the time came, we would rescue him from you."

Judge Hines seemed to accept the Sheik's comments, because she turned away, opened the door and, as with Veeby, found the same two guards barring her way. Judge Hines left the door open and turned around to the Sheik.

"Move them!"

They moved as the Sheik shouted something, but as Judge Hines walked out onto the deck Kelly seemed reluctant to leave the comfort of the lounge. The Judge went out on deck alone.

14

Veeby felt satisfaction as he watched the City of New Boston fade into a misty distance around the Judge's silhouette in the open doorway. Veeby almost laughed watching Kelly's abject submission as the Judge glared from the Sheik to him. She slammed the door angrily.

Veeby realized that he was enjoying himself. This Kamsi had invited him out to the yacht to corral him into salvaging their desperate efforts. To get information and help out of a scientist and in the process had confessed to manipulating the UR legal system. That damn Kamsi Sheik had admitted to getting Kelly a 'pass' to easy prison time and facilitating a prison break. And, to top it off, Kamsi wine tasted pretty good too.

Besides, who knew how much more dirt could be shoveled at this Kamsi prince if he caused trouble? Veeby lifted his half-finished drink and the Sheik refilled it.

"I'm enjoying the cruise," Veeby said pleasantly.

The Sheik gave orders and Veeby felt the ship respond as the rudder went hard over turning the ship around. In thirty minutes Veeby would be back on the shore and then, maybe, a light dinner at home. He always enjoyed dealing with politicians. Even foreigners. He swigged his drink in one gulp.

The poor hapless saps! Rich Kamsi or broke Irish. No difference. They were all on the take. But of course so was he. His smile faded. Yes, he had come out to the yacht to meet with this Kamsi with the hope that he could make some money. Oh well. Better luck next time.

He looked up to see the Judge entering the lounge again and one of the servants stepped inside with her. Kelly, who had not moved, whispered something to the servant. Veeby wondered if Kelly spoke Arabic.

Humph! Kelly seemed to have found his social status. Consorting with Kamsi servants. Across from him the Sheik got up and shoved his chair aside. The Sheik bent over to Veeby whispering conspiratorially, "I have something for you."

Kasan looked up to see the Judge, Kelly and the guard watching. When the Sheik stood up the Judge turned to the doorway, reached in and pulled Kelly outside with her.

"Get out here. You're with me. Not them!"

When the door closed Veeby saw that the Sheik wanted to talk. The Sheik Veeby went into a huddle with him. The Sheik pulled an envelope out of a pocket producing a letter. Veeby took the offered letter that was written on official Department of Justice stationary. It was a job offer to become the new Superintendent of Prisons; a post that included substantial increases in pay and perks, including oversight supervision of the new Remnant facility in the State of Nevada.

"Well," Veeby smiled at Sheik Kasan. "This is not wasted time after all. My wife will be delighted. She's always wanted to see the West."

"You are a practical man," the Sheik said approvingly.

"Yes I am," Veeby agreed. "Besides I hate wearing that damn lead diaper."

"Diaper?"

"Nothing," Veeby said. "It's not important." Veeby put the letter in his pocket.

"Remember," the Sheik said, "Ebbtide will be back in New England in two or three days. We can't let him get away."

Veeby patted the letter, mumbled something about confirmation by the Remnant Senate, received assurances, and warmly shook the Sheik's hand. It was, as far as both men were concerned, a most profitable hour.

An hour and forty-five minutes later in one of Old Boston's most famous Irish pubs Judge Hines sat in a booth across from Frank O'Hannon. Diagonally across from her, at a small cramped table, Kelly sat with a uniformed cop assigned the duty of watching him.

The Judge breathed a deep sigh of satisfaction when Kelly ventured a look in her direction. When she caught him looking, she scowled and Kelly hung his head. He was showing all the right signs. He was at her mercy and knew it.

Good, she thought, one down and one to go. The main prize, Jacob, would soon be in her reach. Now at least she understood how Jacob had engineered the breakout.

She had to admit it was clever. All he had to do was maneuver the legal system to transfer him and Kelly to Stonegate and rely on his Kamsi friends to buy off that crook Veeby. She had to give Jacob credit. The breakout had the simplicity of genius.

Those damn Kamsi were getting their way by buying off officials like the esteemed Warden of Stonegate, but she had plans of her own.

Sure, they could use Jacob for a while. They could squeeze work out of him. Fine. She would get him in time, but more than that, oh so much more, she would get the goods on the Kamsi and their hireling the good Doctor Donald Veeby.

Now it was her turn to use the system. She would make some real money on this. Now, for the first time, she had a real score. Not scratching along on a Judge's pay or getting money out of deadbeats.

Kelly was back in the bag and before too long she'd have Jacob. And, she'd be tinker-damned to turn him over to the Government. She had stonewalled them once and would do it again. This time her wayward prisoner, when he was brought back kicking and screaming, would mow the grass with his teeth! He'd carry that damn lawn mower on his back while he was down on all fours nibbling the grass!

She let herself enjoy the luxury of a smile thinking about what it would cost the Kamsi and Jacob. She looked over the table to see Frank smiling back at her. He seemed to think her expression of happiness was directed at him.

"I need a favor Frank."

"Likewise," he said. His deep gravel voice hadn't changed in years. "Jennifer is doing time upstate. She got nailed on that hijacking thing."

The Judge nodded. Good. Always trade favors rather than pay money. Let the system carry the baggage. That bitch Jennifer was in and out of institutions. Fancied herself a smart career criminal. Well at least she was learning a trade folding sheets.

"Sure. I'll see what I can do."

Frank waited patiently. His big beefy hands rested on the table. The Judge wondered how many men he'd strangled with those massive paws. She leaned closer and Frank bowed his shaggy head to listen.

"I need a snatch job," she whispered.

"Sounds reasonable," he whispered back, "Who?"

"You remember Jacob?"

Frank's grizzled old face looked like it would crack into a million pieces as it forced his lips up into a broad smile. His tired old eyes shone with an inner light making them hard as diamonds.

"Sure," he said softly. "Sure. That punk ran with the Israeli mob. Fancied himself a wise guy." The Judge reached across the table taking one of his huge hands in hers.

"Nothing personal Sweetie," she purred, "but I seem to remember that he had you and Jennifer on the ropes real good with those photos of Bobo hacking up Jennifer's husband."

"Her lover," Frank whispered. "We don't do our own family."

"Yeah well. Those photos were good."

"He muscled in on my family," Frank said.

"That pimp actually figured he could out-smart me with the Israelis. For a guy that supposed to be so smart he's stupid."

"Don't I know it!" the Judge laughed.

She untangled her hands from his and brought her hands up to his face, gently holding his head in her hands.

"It's good to see you Frank. And, by the way this must be a first class job. He's valuable."

"Payoff?" Frank asked.

"Big. Maybe in the millions. He's got the Kamsi over some kind of barrel." Her face lit up into a mutual laugh with Frank.

"Who's the competition?" Frank asked. "If Kamsi are involved can the Government be far behind? All this country thinks about is energy. Where are the Feds?"

"They're asleep at the switch," the Judge said. "I had Jacob at my place for almost two years. Not a peep out of the Feds. I don't think they're interested or if they are," she shrugged, "They're too stupid to be in the game."

"Anyone else?" Frank asked. "Any other players?"

The Judge adjusted her posture on the bench, thought for an instant and shook her head. "The Kamsi have the Warden over at Stonegate in their pocket."

"And you know this how?"

"Because they tried to recruit me two hours ago. A Doctor Veeby, Donald Veeby, the Warden out there at that radioactive pesthole was on the ship."

"Can he put together a crew?"

"I doubt it," she answered. "He's a PhD. He's not a physical type. He reminds me of Jacob. He's in over his head. He's drowning before he knows he's in deep water."

Frank laughed appreciatively. "The world is filled with'em."

"Yeah. Sheep that think they're lions."

Her gaze strayed back across the room to Kelly. A waiter was placing a tall brew down in front him. The Judge waited until Kelly lifted the brew to his thirsty mouth.

"Kelly!" She straightened her shoulders. "Nobody told you to order!"

She smiled back at Frank, who was looking over his shoulder. When Kelly submissively put the mug back on the table Frank whistled through his lips.

"You're tough."

She reached over affectionately pinched his cheek.

"Don't you forget it Sweetie." She handed a note across the table. "Here's the facts," she said. "When and where the mark will be." Frank pocketed the note.

"Now here's the deal," she said confidentially,
"Jacob is to be held someplace safe."
Frank nodded. "No problem."

"I will make like I'm helping the Kamsi. You know. My making use of my underworld connections." She grinned as Frank's face blossomed into a huge grin.

"Double-cross?"
"Hey I learned from the best. Didn't I?" She leaned back, nodding slightly while holding his gaze. "A big payoff Frank. Huge. The Kamsi are desperate. They'll pay through their teeth to get Jacob."

Frank folded the note, tucked it into his breast pocket, and slipped out of the booth. The Judge noticed that he was still a very strong man that moved effortlessly. She reminisced about Frank as he wandered out stopping here and there to acknowledge long time members of the Irish mob. At the other table the uniformed cop was having a beer.

Drinking on duty was still a violation of regulations, but like everything else there was no enforcement. Kelly reached for his beer occasionally, but withdrew his hand. The Judge smiled. That was one puppy she controlled, but now it was time to fit him to a new leash.

"Kelly!"
He slid into the bench vacated by Frank O'Hannon. The Judge paid carefully attention to his eyes. He didn't make eye contact with her and that was good. She could make his life a living hell and he knew it.

"Tell me about Stonegate," she said.
"It's hell on earth," Kelly said. "No lights or running water. Everybody lives in mud. When the wind blows you smell dead animals."

"No lights?"
"Not in the cells. They have a cafeteria. They have, you know, restrooms, but the cells are caves. It's dark all the time. The

prison is carved out of a hill. On top of the dirt it's nothing but radioactive junk."

"And you and Jacob found the way out?"

"With the Warden. I swear it Judge. We couldn't have got out without his help. You need a map. You know. Otherwise, I mean, you're dead with the deer. And believe me nobody is going to scoop you up. You lay where you drop."

The Judge noticed the uniformed cop reaching for Kelly's beer. "Go and get your beer Kelly."

For an instant he looked up. His expression of gratitude looked like puppy love. Kelly went over, grabbed his beer, and brought the mug over, but sat waiting. When she nodded he took a long satisfying draught. After wiping his beer moustache he sat the mug aside.

"I've decided to give you a chance," the Judge said. "You do this and I'll see what I can do for you. Got it?"

"Yes Judge. Thank you."

"I'm going to give you some room. You know that Jacob is still on the loose don't you? You help me with this and I'll help you."

Kelly finished his beer. A waiter walked by and Kelly signaled for another. He raised an eyebrow, checking with the Judge, who nodded. Kelly was appreciative. His every move, every expression, conveyed submission. Judge Hines couldn't remember when she had enjoyed herself more.

"I'd rather work for you then the Kamsi anyway."

The Judge reached over poking at his chest. "You're not working for me Kelly. No money will change hands, but if you dodge out on me you better not be caught twice! Understand?"

"Yes Judge."

"One more thing." She lifted a pair of insoles out of her handbag. "These are special rubber. Slid them into your shoes. They'll protect you from the grid."

Kelly took them with both hands as if reaching for the Crown Jewels of England. He clutched the rubber insoles to his chest; and, while she couldn't be sure it looked as if Kelly was ready to shed tears of joy.

As he swung around, untying his aging work boots the Judge wondered how prisoners could be so thick headed. Didn't they realize that, even if the sidewalks were electrified, (and they weren't); the power outages were so frequent it didn't matter? Besides a little rubber under their feet would insulate them from any electricity in the sidewalk. How could they not figure it out?

Well, of course, that psychologist had explained it at a seminar for law enforcement. She had called it 'aversive control.' "One good hotfoot," she explained, "and the men will be completely conditioned to fear the system. Yes, even when the power fails, they won't risk escape."

While Kelly fumbled with his shoes, making noises of abject appreciation, the Judge idly wondered what means of control they used in the women's prison system.

The Judge smiled again. She let herself enjoy the moment. God, she thought, without the never-ending energy collapse, without the Kamsi throwing money around and without Solbean or Jacob, she wouldn't have Kelly to terrorize.

Four miserable days later, early in the morning, unshaven, dirty, tired and completely disgusted Jacob stepped off an airplane from the flight from Wolfenhausen, Prussia onto the tarmac at the airport in the Democracy of New Maine.

He rubbed his sore wrists where the handcuffs had chaffed his skin. Hancock and Suboy followed him down the ramp. In ten horrible days of on again, off again air travel they hadn't been more than five feet from him and they smelled like pigs. The morning was overcast and hot and the air seemed to accentuate their body order.

"Over there," Hancock said pointing to a small brick structure. They entered a rundown building with no electricity that passed for the airport terminal building. Jacob noted the many other aircraft parked around the facility. Some had rust streaking down once gleaming aluminum.

"This way." Hancock led them to a small room where a cardboard table was set up in one corner with two large candles both of which were emitting large quantities of blue smoke making the cramped room dense with a noxious haze.

Deep in the room shadows moved swirling stagnant smoke into eddies. Jacob looked over his shoulder to see Warden Veeby coming out of the dirty air. He walked forward silently carrying a large leather briefcase that he put on the cardboard table.

Jacob's heart froze as his mind conjured up visions of hot summer days in New Boston with special lawn mowers or dead animals on burned grass. "Please god," he whimpered.

"Doctor Ebbtide?"

On hearing Warden Veeby's voice Jacob's knees got weak. His mind conjured up visions of digging a nice deep hole in which he could spend the rest of his life.

In that terrible instant Jacob couldn't figure out which fate he dreaded most: lawn mowers or a dark holes.

"Who's in charge here?" Hancock asked loudly.

"I am," Veeby said.

"You get the prisoner?" Hancock asked, but before either could answer Veeby pulled transfer papers out of his briefcase.

Jacob slumped into a seat as Veeby came forward to show his identification and sign the transfer papers.

"Here's the papers," he said.

Hancock took the offered papers, hunched down over the candles, showing them to Suboy. They both read over the documents and seemed satisfied.

In an instant the papers were signed. Jacob watched Veeby take the papers from Hancock and then turned his attention to Jacob. "Just like that," Veeby said.

"Right," Hancock pocketed the papers and both men hurried out of the room. Once outside they picked up a running argument about how to get back to the Kingdom.

"We head through Prussia," one of them said.

"They seem to have fuel."

"Ten days!" the other shouted. "Screw it. Let's get to a hotel."

"For a cold shower?"

Their running argument faded. Jacob sat at the table with his head over. He was at that moment a beaten man. He actually envied those two clowns, because they had jobs and someplace to go.

He had spent his life working to be good at something. At first trying to be top-notch spy, but like everything else he tried that had turned against him.

Somehow situations or people always out-smarted him. He couldn't seem to catch the rising tide. More than anything at this moment he was confused. How can some people try so damn hard and keep failing? Every time they turn around something else is going wrong in their lives. Is it all bad luck?

He heard Veeby come up standing behind him. He had the fleeting impression that he could make one last, desperate, run for it. Hancock and Suboy, his two bodyguards, were gone. Or were they?

He wasn't handcuffed, but that didn't mean they weren't waiting for him right outside the door. Oh no, not this time he thought. Not this time! That was exactly the sort of impulse that got him into trouble.

Besides he was still hobbled from his encounter with that nail. That nail, that stupid fat Kamsi guard seemed to sum up his life. Just when he couldn't take any more, just when he decided to make a run for it, his sheepskin boot mashes down on a nail that's standing there just waiting for him. So, how far would he get on one good foot?

Jacob sat with Veeby standing behind him saying nothing. Jacob sat and tried to bring his life into focus. A wave of remorse swept over him. He was beaten. He knew it. He tried desperately to make sense of things and to identify where he had gone wrong. What was his one terrible mistake? What sin had he committed that was so terrible?

That deal with the satellites? If not him then another scientist would have wrestled control away from the

Americans. That beef with Hasham? If not him, not some other westerner would pass gas at the wrong moment and throw the Kamsi legal system into convulsions.

Jacob tried to remember the one place, the one lapse in judgment or the one little slip of attitude that had brought him to this. In the United Remnant and later in Israel he became a scientist and a good one at that. He was dedicated to science and to Israel. He had risen quickly to the top of his profession as a world-renowned genetic scientist.

He had tried to protect Israel by seizing control of two satellites that threatened Solbean production; and, in that one act of liberation for Israel he had turned his own life upside down. Even the President of Israel had turned against him. *But why?*

Now he was a felon. He was, in fact, an escaped felon that was now recaptured. He was angry thinking about all that education going to waste pushing lawn mowers. Worse than that they were lawn mowers with brakes. Now, he was dogged tried. Veeby leaned down to him.

He put his hands on Jacob's shoulders. "I'm sorry," he whispered.

It took all of Jacob's strength to reach up flipping Veeby's hands off him. He sat feeling almost as if he didn't have the strength or the will to breathe.

"I did some checking," Veeby said. "Everything is for sale. Your wife never reported your arrest in New Boston. She kept you for herself. That's why the Feds never came for you. They didn't know you were in the country."

"That's what Kasan said." This was confirmation. He had gotten into the spy habit of always confirming information with a second source.

"They don't want me."

"I think they do," Veeby said, "But you were caught between your wife, the Arabs and the Israelis."

"My wife I understand," Jacob said. "Ditto the Arabs, but Leviman leaves me…" His tears flowed down to the table. He had spent the last year wondering why the Government had shown no interest in him, but especially wondering why Leviman wanted him dead.

Gladys ran her own little serfdom and he was one of the serfs. That was terrible, but to be thought a criminal in Israel was insufferable.

For a long moment Veeby kept his distance. Jacob wept. He let tears flow. Slowly, with trembling hands, he wiped his eyes. Veeby fished around in his large leather briefcase and pulled out a sheath of papers. "First," he said, "The Feds paid your tab with the Kingdom. We have first debs on you." He looked over the table at Jacob. "Now your obligation is to us."

"Don't hold your breath," Jacob whispered. "Write me off as a bad debt now and save yourself time and accounting bills."

"You are a bad debt," Veeby said, "Maybe that's why the Government wants you behind bars."

Jacob put his head down on the table and moved it back and forth. The cold, cheap smelly plastic surface felt good against his hot dirty forehead. Maybe he was getting old. Maybe the world was too tough for him.

People like Veeby thrived. They reached out and grabbed success. The brass ring was within their grasp and they took it. Veeby, like Gladys, had power and authority.

He thought about Gladys and wondered that, maybe, by her thinking the idea of being trapped in a cell with metal pellets in his feet, lying on his back screaming or pushing a lawn mower in slave work, didn't qualify as hard time. What the hell, he thought, even slaves didn't have to repay their masters the expense of their upkeep.

He had worked hard to achieve what? Now he was captive again. Somehow, in spite of all his work and education, life had conspired to bring him to this. And, if that wasn't enough he knew he still had every chance of becoming a menial slave again to his ex-wife. Jacob closed his eyes whimpering. He had no pride left. He didn't care if Veeby saw him cry.

"I really tried to help you," Veeby said softly.

Jacob raised his head. His mouth fell open. Emotionally drained, his brain was far too tired to talk. He couldn't comprehend Veeby's effrontery. Was he naturally sadistic? *He*, the Warden of Stonegate, wanted to help *him*?

Veeby walked around to the other side of the table. Jacob opened his bloodshot eyes and watched as smoke swirled around Veeby's face. He seemed self-satisfied.

"I've been promoted," Veeby said. "From the Bureau of Prisons to the Federal Remnant Justice Department." He lifted a small woman's compact out of the briefcase.

He handed it to Jacob. He took the small polished mirror and looked at his reflection. A grizzled old man stared back at him in the glass. Gray hair, dark brown leathery skin and a gray beard made his eyes more pronounced.

"You look good," Veeby said. "You're brown as a nut."

"I am a nut."

Veeby smiled happily. He gestured to the compact. He took the mirror out of Jacob's hand and sat for a moment admiring Jacob's suntan. Then he dropped the mirror back in the briefcase.

Veeby reached over grabbing Jacob's arm and squeezed his muscles. "Solid," he said. Jacob found the strength to push his arm away.

"I'm a not cattle!"

"No?" Veeby relied. "Maybe not, but you'll be traveling by the pound from now on."

Jacob remembered his last night in the Kingdom. Every muscle in his body ached for that little bed, for the robes, for the sandals, but that he knew was not to be his fate.

"So you win eh?" he asked. Veeby seemed confused.

"Say what?"

"We are going back to Stonegate?"

"No. I've been promoted," Veeby repeated. "Although your ex really kicked up a stink about your escape." He smirked at Jacob. "She was fit-to-be-tied when I got promoted."

"No Stonegate?" Jacob asked.

He knew now what a side of beef would feel like if it could understand while waiting to be auctioned off. Veeby put the briefcase on the floor and leaned over the table close to Jacob.

"You know, Doctor, there really are people who wish you well." His gaze went to the door. "Hines is not one of them."

"No Stonegate?"

Veeby's smirk opened into a wide grin and pushed a paper at him. Jacob read down the Prisoner Transfer Document. Most of the paragraphs were routine legalese. Then, toward the bottom, it gave his destination.

"The State of Nevada? What the hell is in *Nevada*?"

"It's the Federal version of Stonegate," Veeby said still grinning. "Except it's real hot. Soon," he added, "You'll know what hell is like."

"Why?" Jacob asked. "The government still doesn't want me? They know I'm in the country?"

"Sure. They know," Veeby said. "They don't care. You over rate yourself Buddy."

When Jacob hung his head again fighting back another attack of tears Veeby lifted his briefcase on the table. He nudged Jacob.

"Hey," he said poking him harder, "So you'll know. You're damn Kamsi friends tried to buy me. The bastards actually threatened me. Yeah. The bastards thought they could force me to hand you over."

"Over to Gladys?" Jacob asked not comprehending.

"Over to another organization working for the Kamsi. A fake prison. They'd give you a song and dance about not being able to clear you with the law. They would encourage you to stay low. They would make a safe area for you. The Kamsi still want their teeth in your flesh. I want you to know. That's all."

Jacob got slowly to his feet, but kept his hands down on the table. The days in the airplane cramped his back. He hurt all over.

"You deal with the Kamsi too?" Jacob asked.

"They got me this promotion," Veeby whispered, "but I hate the bastards for shipping that rotten shit to us! Screw'em! I may be crooked, but I'm not a traitor! Screw'em! The same goes for your damn wife!"

"I'm not married," Jacob said.

"You know what I mean!" Veeby nudged Jacob away from the table. "Let's go!"

Jacob's tried mind was shutting down. Sheik Kasan wanted him in a fake prison? The Kamsi wanted Veeby to deliver him over to others? What others? Did it matter? Jacob groaned. A fake prison working for the Kamsi would beat Stonegate by a New York mile, but even that was now denied him. Veeby reached over and nudged him harder.

"Let's go," he said again. "We've got three miserable days ahead of us if they can find fuel for the aircraft."

When Veeby motioned Jacob up he realized he couldn't move. His body was exhausted, but not from lack of sleep or want of food. More essentially his mind was exhausted. Stress pulled the life out of him.

Veeby sat down watching as Jacob put his head down. Jacob his kept eyes open and his mind became suddenly alert, although he still could not muster the energy to move. He simply couldn't find the strength to start another long trek to some hellhole in Nevada.

Veeby didn't seemed to be in a hurry either, because after a few minutes of sitting quietly he started talking again rambling on about Judge Hines.

"You know her better than I do," Veeby said, "but I know her too. There was a real flap when you escaped from Stonegate. It was a mess. Hell," he added, "If I'd known you had been *married* to her I wouldn't have touched you or that crud Kelly with a god-damn ten foot pole! You cost me a lot of political favors!"

Jacob blinked his eyes, but didn't raise his head. He felt as if all he wanted to do was stay in the smoky room letting candle light flicker a few feet from his face, watch the flames dance, and forget his troubles. Veeby kept talking.

"See, Ebbtide, your ex filed charges against me for aiding and abetting your escape. We met on some Sheik's yacht and the very next day she filed charges. That was three days ago."

Jacob watched as a gentle breeze from the doorway blew smoke at Veeby. He moved the candles to another part of the table he moistened his mouth and continued talking.

"See?" Veeby added. "Even with all her huffing and puffing I got the promotion. Fast too. I think your ex got the message about my political connections. She won't tangle with me again. Rest assured. My connections are better than hers that's all. She didn't believe for an instant that you and Kelly could wander out, find the right boxcar, and ride out of Stonegate. She knew you had help."

"Now you have a new job?" Jacob asked.

"I got promoted," Veeby repeated. "That's why she's angry."

Jacob kept his head down on the table. So, Veeby won the political power contest. How he won or why he won wasn't Jacob's concern.

Who Veeby was double-crossing the Kamsi or the Judge. It wasn't his concern. Now it would be Nevada. Jacob wished again for the good old days pushing the lawn mower and having Gladys sit out under a beach umbrella sipping lemonade. Veeby leaned over as Jacob tried to speak.

"What?" Veeby asked. "I can't hear you."

Jacob worked moisture into his mouth. "Shoot me."

"No fun in that ol boy," Veeby said. "Come on. Get up. We've got hard days ahead of us. Come on! Move!"

Jacob felt his bones creak as he slowly forced his away from the table. He walked ahead of Veeby out of the terminal building.

Far across the tarmac an aircraft painted in black with white letters, 'State of Nevada,' waited. All four engines were running. Hot exhaust shimmered in the air. The sun blazed.

Long black shadows from long abandoned airport buildings promised a long sweltering lousy day. Sounds of jet engines caused Jacob to look where Veeby pointed.

"There," Veeby pointed. "Over there."

People aimlessly milled around the outside of darkened hangers. Armed security guards moved cautiously between the shuttered airport buildings. Veeby studied the uniformed personnel with a practiced eye.

"See?" he asked. "Even security personnel don't go into those damn buildings. The corridors are all dark. The air inside is foul and who knows what's waiting for them?"

Jacob stood beside Veeby and felt like a dead man standing beside his own grave. His shoes weighed like lead. His right foot blazed in pain. His gaze went instinctively to Veeby's midsection, but there was no diaper. Jacob's eyes blinked in sunlight as his gaze followed Veeby's to the airport buildings.

All of the buildings looked unused. Jacob understood why. With no electricity the heating and air conditioning systems rusted in no time.

Without use moisture seeped into the wiring. Short circuits resulted. The support systems built on cheap oil energy were now scrap.

Standing by one of the buildings a security guard slid a rifle off his shoulder. A bunch of kids rushed out a building door to the guard's right. Two kids bumped the guard.

"Hey!" the rent-a-cop shouted as the kids caught him from the side throwing him to the pavement. The guard rolled, cradled his rifle, aiming at the retreating kids when another larger gang rushed out the same door and kicked the rifle out of the his hands. The kids trampled over the guard. Jacob stood as if shell-shocked at the fast moving band of little thugs.

They were organized. It was the old 'one-two' punch. One group distracted the guard by acting as a decoy. They hit him once and the second bunch caught him from the rear kicking him to the ground. Jacob saw a dark stain form on the guard's back.

"Golf shoes," Jacob said.

Veeby pulled a handgun from a deep pocket and held it casually by his side. His attention went to a lone man working his way from building to building by ducking into doorways.

He seemed to be trying to get behind the kids in much the same way that the kids were out-maneuvering the guards. Jacob could just make out a man moving fast from door to door.

"Recognize him?" Veeby asked.

"No."

"I think that's Kelly."

Jacob rubbed his tired eyes, but the light and glare on his tired eyes was too much. The man moved fast and seemed to know what he was doing, but was it Kelly?

"Why would Kelly be here?" Jacob asked.

"That bitch wife of yours captured him."

Jacob stood next to Veeby trying to figure out what the hell was going on. The airport tarmac was alive with men. Gangs of kids attacked airports guards while other guards took shots at the kids. Anything wearing a uniform seemed to be taking most of the blows. The bullets didn't seem to hit anything.

The action was fluid and fast. It moved from the tarmac to the buildings, in and out, and back to the open areas again with the sounds of gunfire reverberating off of tin walls.

Kelly ducked as the kids, covering their action, fired shots at him and then in quick order returned fire. When the guard dropped on the tarmac another guard came around a building and started shooting at both Kelly and the kids. Kelly ducked into another door while the kids ran for a wire fence a couple hundred feet distant.

A small group of travelers ran out from the other side of the building, out across the nearest runway, heading out to the black aircraft.

"There!" Veeby pointed, "We have to make a run for it."

"I can't," Jacob said, "Honest."

A light aircraft landed on a near runway. Jacob looked up to see an electric light beaming from the aircraft control tower. They were using light signals to direct air traffic and using their limited energy resources effectively.

Jacob's tired mind worked on the problems of air traffic control with plummeting energy resources. His mind was closing down and distracting itself with any random thought rather than facing the reality that he was going west.

A metal object, dark blue and polished, entered his field of view. Jacob's eyes focused on a handgun.

"I insist," a masked man said, "or I'll leave you dead right here on the tarmac." Jacob lifted his bad foot making an effort to run when a man's voice behind them startled him. "No you won't."

15

The man's handgun hit the pavement. His weapon seemed to fall in slow motion. He raised his hands in a gesture of surrender. The masked man hit the ground.

"You," the other man said with the gun said. "Get over there."

A noise caused Jacob to look at the kids. They had changed directions. They ran toward them now, but fanned out like wolves circling a caribou. Jacob's heart jumped.

Veeby grunted as loud claps of two shots echoed between the buildings. Jacob looked to the two gunmen. One was already on the ground. The kids rushed over, stomping on him. They came back fast in Veeby's direction.

With a gasp Jacob realized that Veeby had retrieved a gun from the ground. Both two masked men were on the ground, but still alive. He knew, because blood spurted out of neat little holes in their necks. God, Jacob thought, those little thugs can really shoot.

"Move it," one of the kids yelled.

Veeby grabbed Jacob's coat sleeve and shoved him toward the waiting aircraft. The kids bunched around them, running with them, yelling, firing random shots into the air. Jacob forgot to breathe, but ran with the kids bunched up around to protect him and Veeby all the way out across the runways to the waiting aircraft.

Once at the aircraft Veeby pushed Jacob to a ladder. "Go," he commanded. As Jacob climbed saw Veeby take a small instrument out of his pocket and toss it to the kid.

The kid grabbed the instrument. "Yeah Big Fish," he said. "See ya."

Veeby grabbed hold of the ladder and started climbing. Halfway up the ladder he shouted down, "I'll put in a good word for ya."

"So long Big Fish," the kid yelled and ran off at a fast trot, firing randomly to discourage anyone else from interfering. Standing in the open aircraft doorway Jacob had a clear line of sight of the gang and the security guards struggling on the ground. He didn't know who the other men were, but they weren't kids. A guard looked up to see the kids coming at them again. The kids ran over him for a third time.

The situation around the buildings was chaotic. The kids split up into two groups again. Both groups ran around like military units while the man Veeby thought was Kelly weaved and dodged like a man that knew he better run if he wanted to keep breathing. Both groups took shots at him. Jacob turned away from the door as a steward took his elbow.

"They are my backup," Veeby said. "Your former wife knew you would be here."

"My god," Jacob groaned in sudden understanding. Gladys was running her own game. As Jacob hobbled down the aircraft isle he heard more gunshots outside the aircraft. Sounds of gunfire echoed off the metal buildings and reverberated through the aircraft interior. Once in their seats Jacob looked over to Veeby. "Those kids are something."

"They are the future," Veeby answered. "They can survive on next to nothing. They have no illusions."

"You bought them?"

"Yeah, well, I do them favors. You know. Even as a chemist with the Department of Energy I knew the kids. That was in Chevy Chase, Maryland where I used to life."

"I suppose it's useful to know people like that."

"It can be," Veeby said. "But you have to have something to offer them. Friendship doesn't cut it."

Jacob nodded. Yes he knew. The Warden of Stonegate was a good guy to know. He could make the revolving door of justice revolve faster. A crew could be back out on the street after a good night's rest, a hot meal, and a clean shower. Leave it to the justice system to get things backward. Scarce resources spent in all the wrong places. Democracy in action.

"My wife the mobster" Jacob said closing his eyes.

"She knows many of the same people you and I do and some we don't." Jacob's tired body wanted sleep. His brain barely heard Veeby's rely.

"Never forget she was a mob lawyer," Veeby said. "I found out about her when I did some checking. In today's world it doesn't pay to be stupid. Mobster or mob lawyer. It's a subtle difference."

"My ex the mobster," Jacob said sleepily.

He was relieved to hear jet engines rev up. The aircraft moved out onto the runway and Jacob was asleep by the time it lumbered into the air.

*   *

Kelly was desperate. His effort to get Ebbtide away from Veeby was a complete fiasco. With all the shooting going on all he could do was run around on the periphery and try not to get killed. Those hoods that O'Hannon sent were dead in seconds.

Why was he the only player? Kelly understood the Judge had a deal with Frank O'Hannon, but he also understood that if Frank's men got Jacob she'd have to split Kamsi ransom money with the O'Hannon mob. Therein was the rub. The Judge didn't like sharing especially when it came to sharing money. Getting Jacob was Kelly's ticket to a better life.

Kelly knew that he was her triple-cross. He also knew he'd be hell bent if he went back to New Boston without Jacob in tow. A gunshot pinged by his head.

Those damn kids! Those roving gangs! They had no respect for adults! The little bastards could shoot! Kelly ducked into a building and tried to figure things out. What was his best maneuver?

He couldn't go back to New Boston. Not yet. He couldn't walk back to the Judge. Not without Ebbtide. Even killing Veeby would have gotten him some points with the Judge, but he couldn't even get a shot. Those damn kids!

He took a piece of paper out of his pocket. New Haven in the State of Connecticut had decided to stay with the United Remnant. His rental car was good for travel to New Haven.

He wouldn't cross any national boundaries requiring a security pass or payment of a fee. Ebbtide had a friend there. His name was Zada. George Zada. Maybe he could help. Kelly had his address, but had got caught before making contact.

Ebbtide said he had trouble with authority. Well, maybe Zada would be willing to help him get to one of the other splinter countries. Maybe he could help with a new I.D. Besides what other recourse did he have? Maybe Ebbtide knew something.

Maybe Zada had connections of his own and could find out where, exactly, Ebbtide would be imprisoned. Or, maybe just a new ID for himself would be the ticket. But he couldn't get caught again.

If he did he'd have to find some way to score points with the Judge. He couldn't go back empty-handed. That was out of the question. But still, there was another problem.

Getting to New Haven was a four-hour drive around New Boston. It would look like he was making a run for it. He would have to chance it. Kelly made his way to his rental car, but stood in the deserted parking lot staring.

Kelly walked closer daring not to believe what his own lying eyes were telling him. The vehicle's wheels had been stolen. The vehicle sat on its axles. Those damn kids! Kelly approached the car stunned by his repeated bad luck and put his head down on the metal roof. With his eyes closed he rested his head on his arms wondering what else could go wrong. Those damn kids! Then he felt the touch of cold steel on his neck.

"Hand it over," a kid's voice said.

Kelly reached into his jacket and slowly slipped a police special .38 out of the holster. He handed it to the kid. It was the Judge's personal handgun. She was going to be *pissed*.

"Tanks," the kid said.

He sounded like a ten year old. Kelly froze. The young ones were the most deadly. They'd shoot at the slightest provocation or no provocation.

"Tanks to you," Kelly mumbled, but immediately cold steel tickled his neck.

"You got cash?"

"Hip pocket," Kelly said.

Without missing a beat the kid used his other hand and neatly plucked Kelly's wallet. Kelly heard a truck drive up and brake to a stop.

The kid jumped into the truck with Kelly's wallet and handgun. As the truck departed Kelly straightened up taking a deep breath.

Looking around fugitively he wondered about the kids, but also about other spies the Judge might have watching him. The parking lot was empty. In the far distance, by the airport tower, the security guards were still lying on the tarmac.

Those two rent-a-cops called to one another like two dying ducks in winter.

"Now," Kelly said thinking and wrestling with his new situation. First, he'd check for tools in the trunk. He still had the car keys. The kid hadn't bothered with the tools. Who would use them? He needed wheels that would fit his rental car. Kelly's attention went to the problem of getting new wheels.

After breaking into another vehicle Kelly used the owner's own car jack to hoist the vehicle and then he went down the line of cars, getting jacks and swiping wheels. Better to be safe than sorry, he told himself as he swiped a second set of wheels and piled them in the back seat. Besides he might be able to sell the extra wheels for a little cash.

Two hours after stealing wheels and more four hours on the road Kelly turned off the vehicle ignition. The New Haven neighborhood had once been a middle class neighborhood that, like so many others, had fallen on hard times.

The rundown dwellings were obviously rentals. Multiple addresses advertised the fact that the original homeowners couldn't pay their taxes and had to sublet their property. Potted plants, here and there, were a pathetic effort to beautify what had once been a pretty neighborhood.

Up and down the block lawns sprouted large brown areas where dirt prevailed over grass. One of the houses, however, setting on a rear lot was surrounded with lush vegetation. Whoever lived there used a lot of water. It looked like the jungle had recaptured the whole house and yard.

Masses of yellow and green vines swarmed over power and telephone poles and a lean-to type front porch. Even the walkway, concrete slabs, leading up to the front door was a thick mat of tangled vegetation. When wind changed coming off the property Kelly imagined that what the Amazon must smell like: heavy, lush and moist as if a giant oven had opened to let the heat out.

Kelly let his gaze roam over the rest of the street. Down the block people congregated while men rushed out leaning in car windows. Drug dealing.

Police were nowhere around. Kelly pulled the paper out of his pocket. The address given him by information, based on the phone number, could be any one of these units. He decided to knock on doors.

The first two had no idea who the Doctor Zada was, but the third was a pleasant surprise as a very attractive woman came to the door.

"Do you know Doctor Zada?" she asked pleasantly.

"My business with the Doctor is private," Kelly explained, "but also rushed. I have business with him."

"Where are you staying?"

Kelly unclipped his cell phone from his belt and held it up. She made a note of the number. "He'll call you shortly."

By the time Kelly got back his car his cell phone rang.

"Mr. Kelly?" a man asked. In response to Kelly affirmative reply the man said, "What business are you in?"

"I'm in the worried business," Kelly said. "I understand from a mutual friend whose first name is Jacob that you might be in a position to help me."

"I've known many people named Jacob. It's a common name where I am from. What line of work does he do?"

"He's a scientist. He works in genetics mostly in plants."
When Kelly went to say more, but the man stopped him.
"Yes. I know that man."
He seemed to be enjoying the conversation. "Now that Jacob
I remember is a tough fellow. Big and rugged. Always getting
into trouble."

"You've got part of it right," Kelly said, "All except the big
and rugged. I'm big and rugged. He isn't."

"Aah," the man replied then laughed, "And yet it's the big
and rugged *you* that's worried."
"I think you've got the situation. Can you help?"

"That depends on what you want. Perhaps we could meet.
"I'll tell you about a place. It's small and private. Be there at
eight o'clock."

That evening Kelly met Doctor Zada. He was a small man
not unlike Jacob except for a full-face bread. They sat at a
table in an out of the way location.

"So," Dr. Zada said. "You are worried?"
"Yes," Kelly replied.

"And the source of your worry?"
"I have a Judge who got my number. She's made me one of
her personal projects. See, usually if a guy breaks out of jail he
can rely on the system to lose interest, I mean, crime being
what it is today."
"The judge isn't losing interest?"

"I'm her pet project."
"An interesting choice of words," Doctor Zada replied.

He ordered from the menu and indicated for Kelly to do
likewise. They had a big meal, spoke at length about Jacob,
and about the world situation.

Kelly learned that Jacob and George Zada and his brother
Emil went way back to days in Tel Aviv and even as students
in America. All three were geneticists. All were mavericks, but
beyond that their work had taken them in different directions.

The conversation turned to what Kelly really needed. "I
need a favor," he said.

"Any money?" Doctor Zada asked. "Favors cost money. I am not running a charity."

Kelly gulped. He was afraid the conversation would turn to the issue of money.

"I have some merchandise," he said. "Worth something on the black market." When Zada nodded Kelley continued,

"I can only tell you, Doctor, that the source of my worry has also made Jacob one of her pet projects. She is using me to get him. I'm supposed to kill Jacob or kidnap him.

"Or, if all that fails at least get a line of his whereabouts. Frankly, I've got nothing against Jacob. He's my friend, but I'm getting pushed. You see my situation?"

Doctor Zada nodded. "It's unfortunate," he said.

"Yes. The law can be quite intimidating. People with power tend to use their power. You say the source of your worry is a Judge?"

"Yes," Kelly said, "and she has influence."

Zada sat, thinking and finally seemed to come to some decision. "I work in genetics," he said. "I work with insects." When Kelly said nothing, but sat attentively watching Doctor Zada pointed out to the street.

"Merchandise you say?"

"In the car," Kelly said.

The following day about noon Kelly slowly opened the door and ventured into the darkened pub. The stink of stale booze whiffed into his nostrils. A haze of purple smoke swirled around his head. It was good to be back in civilization.

Kelly slumped into a booth and felt a mix of raw fear and shame. How could an ex-cop, a tough man, a guy convicted of multiple homicides be so damn pussy-whipped? How could a man be so terrified of a woman?

"Kelly!"

He looked up to see the Judge entering the dark establishment. She came over with her hands on her hips. She obliviously had heard the news.

"Well! Give me your report!"

"Frank's guys were out-gunned," Kelly said. "There was nothing I could do. I couldn't get Ebbtide away from Veeby."

The Judge stood outside the booth eyeing him with that withering gaze. When he stopped talking she put her hands down on the table.

"Is that it?"

"Those kids," Kelly said. "There had to be fifty of them."

The Judge leaned down closer. Her voice was low.

"I relied on you! You were the backup. We knew that Veeby was in tight with street gangs, but so are you! You used them to set up your own wife!"

"I didn't know *these* kids," Kelly explained. He held his hands up helplessly.

"Did Frank's people move in?" She asked.

"Yes," Kelly replied remembering how fast they had died. "They were dead in four seconds flat. The New Boston Police should have men that can shoot like that, I mean, Judge, they could clean up this town in two days."

"And you?" she asked her voice hard. Kelly cringed. The Judge was pissed and that wasn't so good for his future.

"I'm lucky to be alive," he said. "The kids saw me circling around between the buildings trying to get to Ebbtide and came after me, but I ducked into one of the airport hangers. I was lucky to get away with my life."

"You might not think you're so lucky when I'm finished with you!"

"I'm sorry Judge," Kelly's voice quivered. He dropped his gaze and slowly brought up a leather briefcase

"I managed to get this off of Veeby before he got Ebbtide on the aircraft. Veeby dropped it on the tarmac."

The Judge walked around the booth and sat opposite Kelly. "Well?"

She slid the briefcase over. Her hand rested on it. A waiter came up, but she dismissed him.

"Papers Judge," Kelly said. "His transfer papers. If he's going someplace in Nevada it's all in there including the Department of Justice authorization code."

The Judge straightened up. Her sight focused on the briefcase. Her tongue moved smoothly over her lips. Kelly caught a glimpse of a smile.

"That code, Kelly, is your salvation."

"You mean," he asked humbly, "I did good?"

Hines let herself smile. "That code is the key to getting Jacob back here! With that code I can find him anyplace on the North American continent."

When the waiter returned she nodded to Kelly who grinned his appreciation and ordered a tall beer. The Judge unsnapped the briefcase then looked across the table.

"Go to the bar Kelly!"

He slid to the edge of the bench as the Judge opened the case and reached her hand inside. She reached in, grabbed some paper and simply stopped. Her head lowered slowly to her chest. She made no sound.

At the bar Kelly finished a steak and washed it down with another beer. "Put it on the Judge's tab," he said, "but she likes her naps so don't bother her."

When he went back to the booth he carefully lifted the briefcase, snapped it closed, patted the Judge on the head, and stepped out of the pub into the cool clear light of freedom.

Jacob wondered when was the last time they had repaired the runway with all the noises and racket the airplane made on the pavement. Outside the aircraft building windows whizzed by. Strange, Jacob thought, that the buildings were so close to the runway. The aircraft's wheels bounced and made a racket while the machine came to a stop.

Stepping onto the exit stairway, Jacob blinked in bright sunshine, but also blinked because he realized the runway was right down the middle of the famous Las Vegas Strip.

"Damn shame they had to do this," Veeby said stepping up beside him in the doorway. The major roads leading into the City had been barricaded off. Well, of course, Jacob realized there was no automotive traffic. It would be a crime to let the perfectly good concrete roads go to pot. Then he looked again.

The roads hadn't been built to take the weight, the pounding, of heavy aircraft. The highway surfaces had broken up and been covered over by corrugated metal sheeting.

"Damn shame," he said agreeing with Veeby.

Once off the aircraft they walked along a street with Jacob feeling a tinge of envy. Vegas actually had electricity from Colorado River hydroelectric generators. About half of the buildings they walked by had lights, although none of the massive signs were illuminated. The casinos were dirty, shopworn buildings badly in need of a paint job and basic maintenance.

Poorly dressed patrons opened doors on their way into cool dark interiors. Wonderfully cold air spilled out into the sunbaked sidewalk. Jacob heard laugher and some cheering. A little old time air conditioning went a long way.

Jacob and Veeby walked by another group of rundown street businesses and shabby office structures He saw four camels standing in a tight group in an alley to the side of a casino.

Jacob's heart skipped a beat. For an instant he was back in the Kingdom as white clad men haggled over the price of whipping boys. Behind the casino, a mere block down the alley, Jacob saw nothing but desert. Sand had reclaimed the City. Camels roamed within easy sight of downtown.

"What happened to the rest of it?" Jacob asked.

Although he had never been to Vegas he had seen television and read about Las Vegas. America's playground. Urban sprawl.

A camel stood in the shade of a building, close to the intersection of the street and an alley. It passed gas as they went by. "Filthy animals," Jacob said. "I hate camels."

Veeby tugged on his arm. Jacob stopped as Veeby started talking with two men standing with the camels. Jacob held up his nose and looked askance at the ragged rough-looking men. In the bright sunlight he couldn't see any brands on their forehead, but he wondered if they were prisoners?

"Who are you?" Veeby asked.

"We were sent to escort you out to the complex," one of them said.

"Prove it."

The man took papers out of a camel's saddlebag. Veeby took a quick look and looked again at the animals.

"You're kidding."

Jacob watched Veeby lift a hand and slowly count four camels. A camel did his duty dropping a huge paddy within inches of Jacob's boots.

"They didn't send a car?" Veeby asked. Neither man responded, but when Veeby got close to one of the camels the man cupped his hands. Veeby put a shoe into the waiting hand and got onto his camel.

"Mount up!" the man shouted to Jacob.

"What?" Jacob asked.

"That one is yours," the man said pointing to the pooper.

"What?"

"Get on," he said. "This is how we travel out here."

"Camels are the best way to travel in the desert," the other man said. "You'll learn to appreciate them."

Veeby brushed his hand against his camel's mangy hide. "Damn," he said. "If I'd known about this I'd have…" He let his words trail away.

Jacob stared disbelievingly at the camels. They were huge. They also seemed to have an ill temper. Jacob's skin wanted to sweat, but it was too hot for any part of his body to function. He didn't have a hat.

"Warden," Jacob said, "Even in the Kingdom they have mechanized transportation. Hell, even the Arabs have Centas. Damnit! Even Kamsi have trains!"

"This ain't the Kingdom," Veeby said.

"Let's go!" the men shouted in unison.

One of the men got Jacob into a turban headpiece and helped him onto his mount as the other man handed Veeby a headband.

"We must make time," Veeby instructed, but the men grunted in disgust.

"Camels go slow!"

After a moment of kicking and bucking the four camels lumbered lethargically away from the shelter of the buildings. Jacob was astonished how fast the sounds of humans laughing died away in the wild open spaces of the desert.

"Take it easy," one of the drivers advised. "The desert claims those in a hurry."

One man jockeyed his camel out toward the setting sun. Jacob watched the two men use their sticks on the animals and then turned his weary attention to the land.

This desert was different from the bleak windswept deserts of the Kingdom where incessant wind piled the sand in huge dunes that moved snakelike across the horizon. Here scrub trees with hundreds of vultures sat patiently on thick knurled branches.

Thick black branches and roots littered the desert floor. In the Kingdom the sand was deep yellow, but here it was bleached brown with white and green specks. Behind them the Vegas strip receded in stony silence. Only an intermittent breeze whistled through Jacob's ears. Casinos were still visible along the Strip. Buildings were still in sight as the sun reflected off of desert glass, but he saw no humans.

Hours plodded by at the rate of ornery camels walking. Far behind them a smattering of electric lights gleamed in a deepening afternoon. Huge black birds flapped out of scrub trees as the camels lumbered by and Jacob realized with a jolt that the vultures were circling them. The birds were waiting for a meal.

Jacob rode with his right leg crocked around the camels hump. His camel didn't seem to like him. He mentioned it to one of the others.

"Camels hate everybody," he said. "When it slows down use the whip!"

"My god," Jacob moaned, "I had to come to Las Vegas to see a camel."

In the distance an aircraft winged its way over them heading in the direction of the sleepy casino town. Jacob's camel pulled its head upright and dropped a paddy as the airplane went over.

Miles behind them the aircraft lined up on one of the roads along the Strip for touchdown. Within thirty minutes Jacob saw no trace of Vegas or of civilization. Not even broken shards of glass reflecting in the setting sun.

By late evening they traveled ten miles. It was twilight. The earth was visible, but color had drained out of the air. Jacob saw no sign of a road or a highway or a bridge.

It was as if the only way into Vegas was by air and then only if fuel was available. The only movement in the landscape was big black vultures that clustered in scrub trees and took an unnerving interest in them.

"We travel mostly by night," one of the men said.

"Night is best," said the other. "It's cooler."

Jacob watched Veeby who seemed to be asleep on his camel. The man was amazing. No, Jacob thought, not amazing, but a man at home in the world as he finds it. Veeby was comfortable with life. He was, as far as Jacob was concerned, a lucky man.

"What do I call you?" Jacob asked the leader.

"Call me Der Fuhrer," he hollered back. The other man laughed.

"And you?" Jacob asked the other.

"Attila."

"Adolph and Attila," Jacob said. "I assume you are brothers?"

The two laughed and dug their heels into their camel's sides. Jacob watched them prod their animals as the sun nosed down in the West.

Veeby slumped over in the makeshift saddle and for an instant Jacob thought he would fall off. The Warden steadied himself and looked around at Jacob and dozed off again. The camel train made another five miles before making camp for a few hours sleep.

A campfire was lit. Sleeping bags rolled out. The chow was something hot and soaked in fat. When the meal was finished and the fire still burning bright Veeby brought his sleeping bag over by Jacob.

"Don't get any ideas," he said.

"Where would I go?"

"Well," he looked of in the direction of Vegas.

"You could make it back to town."

When Attila walked up Veeby reached for Jacob's left foot.

"Hey!"

"Pipe down," Attila shouted. "You won't walk nowhere without boots." He laughed. "Hell fire," he said, "You damn tenderfoot! Them flimsy prison boots will get you killed out here what with snakes and all."

Veeby looked down at his footwear. Adolph came over gazing at their boots by the firelight. "Yep," he agreed, "between the snakes and the scorpions you two are road kill."

Attila reached down, taking Jacob's shoes and then gestured for Veeby's. "We'll put'em up in the saddles," he said.

"Be safe up there."

"Always protect the boots," agreed Adolph.

Jacob wondered if Adolph and Attila might not get on the damn camels in the night and ride out leaving them to die in the desert.

Protect the boots? It came as another shock to Jacob to realize that those bozos were free men with jobs and that he was the low man on the totem pole. He was hauled from one part of the world to another like a slave.

Let them save the boots. With his sleeping bag on the good earth, his feet near the fire, Jacob slid inside the folds of his sleeping bag and was asleep almost immediately; because he knew that in a few hours they would be back in the saddle.

They traveled by night, but they also traveled by day. About two o'clock the temperature reached over a hundred degrees. They stopped for a few hours rest. Then they pressed on well into the second night.

Two days later they rode by a sign, 'Yucca Flats.' A mile or two later they entered a narrow pass that opened to a broad flat area stretching out as far as Jacob could see. The sand was orange hued and had the tiny white glasslike particles embedded in it. They went by a sign reading, 'Radioactive area. Danger!'

Jacob looked at the signs and decided that if the radioactivity didn't brother Veeby it wouldn't bother him. Why should he care? He was good as dead anyway.

Toward nightfall on the third day they came up on a group of low concrete bunkers. One lone electric light pole cast a pale white beam out into a vast desert expanse.

"Veeby," Jacob said as the camels came to an automatic stop at the first building, "Do it now."

"What?"

"You know. Do me a favor."

"Do what?"

"You know. Please. End it. I can't take anymore."

Veeby didn't answer. He moved his camel as a big wooden door on the concrete bunker opened. A shaft of bright white light fanned out into the blackness beyond.

"You Ebbtide?" a man asked.

"That he is," Veeby said. "I'm Doctor Veeby."

The man was short, stocky. He nudged his way around the camels and came up to Jacob. "I'm Meehan C. Handy," he said. "I'm a security consultant for the Government. Or, at least that's what you can call me."

"What?" Jacob asked, "Mr. Handy? Handy?" Jacob repeated. "Don't I know you?"

"Maybe," he said. "Or maybe not. Couldn't say. Was it a long time ago?"

"About a year and a half ago," Jacob said. "I called you."

"You're the guy," Handy said as if recollecting.

"That's right. You were working in the Kingdom?"

"No!" Jacob shouted. The shout made his camel buck. It nearly threw Jacob off. "I was stuck in the Hines Prison House!"

"But you just got back from the Middle East? Didn't you?"

"That was before!" Jacob said. After a moment of mutual staring Jacob shrugged. "I give up," he said. "I can't explain it."

"Then don't try," Handy said.

"Just one thing," Jacob said. "Why didn't you make some phone calls?"

"I did," Handy said. "All the connections went back to the Kingdom."

Handy got Jacob off his camel. "Say, I seem to remember a guy named Ebbtide from years ago. That sure sounds familiar. Where'd I know you from?"

"Nowhere."

"What?"

"Never mind him," Veeby said. "He's eccentric."

"I hate eccentrics!" Handy said. "The whole damn world is filled with eccentrics." He pointed to Jacob and then to the entrance. "You! Ebbtide! In there! Get inside."

As Veeby started sliding off his saddle Handy spun around putting a hand up. "No sir. This is a private facility. Boys?" he motioned to the men.

"Mr. Veeby?" Attila said. They were still on their camels. "We ride!"

"You explain about the shoes?" Handy asked.

"Yeah," Adolph said. "Too bad about the prisoner's camel running off that way with his boots. We'll make out the report. Damn shame to lose prisoners that way, but the desert you know."

"It's dangerous," Handy agreed.

Jacob stood by the door half interested and half asleep, while Veeby brought his camel around to square off in front of Handy.

"What about his shoes and camel?" Veeby asked.

"Your man was lost in the desert," Handy said. "You lost him when his camel ran off with his shoes. He's dead. It's real unfortunate."

Veeby looked at Jacob and Jacob nodded to Veeby and lifted a hand. He waved goodbye to Veeby.

"Nobody will believe it," Veeby said determined to stand his ground or rather to sit commandingly on his camel. He struck a pose of a general on his horse. "Nobody will believe it!"

"Make them," Handy replied. "Besides you have witnesses. They'll back you up."

"It's not right!" Veeby shouted.

"Now that you've been promoted," Handy said, "Enjoy it, but don't get confused. Your new office is in Las Vegas. Don't come out here! Don't show yer face out here!"

When Veeby said nothing, Handy came forward and stood near Veeby's camel staring up at him.

"You see, Doctor, you don't have your prisoner. You have not transferred him. Nobody has signed for him! If you lose him and he's not dead," he gestured over to the two camel drivers, "Then you are incompetent! That won't do for a man in your new employment!"

Veeby's camel stepped back as it grunted, passed gas, kicked and spit at Handy all in one practiced motion. Handy jumped away, but kept his gaze on Veeby.

"You're forgetting," Veeby said. "I still have political connections! You won't get away with taking my prisoner!"

"He's not your prisoner now. You lost him."

"I've got big connections!"

"Do you?"

"Yes I do!"

"You've been played Veeby. Enjoy you new employment, but don't make waves."

Veeby's camel walked around so that Veeby's face was out of the building's shadow. Jacob saw his face clearly in bright light. The Warden looked devastated as if he'd lost his best friend.

"Hi ye! We ride!" Adolph suddenly shouted causing Veeby's camel to buck, pass gas, grunt and gallop off in the direction they had come. In the moonlit distance huge black birds flapped their way into a star-filled night sky.

"Viva! Viva Poncho Villa!" shouted Attila. "First we plunder Vegas! We ride on thundering hoofs!"

"First we take the women!" yelled Adolph.

"Hi ye! We rape and plunder!" shouted Attila.

"Life is good!" shouted Adolph.

The drivers got their steeds moving with generous use of their sticks. Veeby's pale face shone clearly in bright moonlight as he looked back at Jacob he seemed to be crying, but maybe, Jacob thought, it was bright fused glass particles reflecting light from radioactive sand.

"I like to see men happy in their work," Handy said waving to them.

Jacob stood on a small concrete slab looking at the desert. He heard the birds rustling and squawking in the trees and saw dark outlines of vultures taking flight as the departing camels passed under twisted branches. In the distance Jacob heard the camel drivers raising hell and Veeby still protesting the whole thing.

"So I'm dead?" Jacob asked Handy as he walked up.

"You perished on the lonely desert Sir. It's a dangerous world out there. Come along now. That's a good man."

"What now?" Jacob asked.

"Now?" Handy repeated. "Now you get to be a scientist again."

"No firing squad?"

"If we wanted you dead, Doctor, we wouldn't have paid the Kamsi to get you back. Sooner or later they would have done our dirty work for us. We know about your trouble in the Kingdom." He hesitated and then added, "But let's talk about that later."

Jacob looked around at the barracks. The ceiling was low and like the rest of the building was made of bare concrete.

Long lines of beds stretched out to a far wall that, Jacob estimated, was a hundred feet long.

"What is this?" Jacob asked. The barracks was empty, but there were dozens of beds.

"Bunkhouse."

"Bunkhouse?" Jacob asked. "You mean this isn't…"

"Stonegate West?" Handy was disgusted. "Hell, this ain't even half way."

As he spoke he moved into the light. For the first time Jacob got a good look at Mr. Handy. His forehead had raised spots; tiny areas that looked like burn marks that hadn't healed. Handy caught him looking.

"Brands," he said his temper flaring. "If I ever find that double-crossing Puller I'll skin him alive!"

Handy turned around to get Jacob's face into the light. "Engineer?" he asked. "Third floor?"

"Yeah."

"Funny," Handy said, "I always knew you'd wind up behind bars!

"But you got the brands."

Handy came up close confronting Jacob, "Yeah, but you got the hotfoot! It's all in your police record!"

Jacob shied away from Handy. He looked around the barracks. He had many questions, but decided that to let the situation unfold. Obviously, someone had gone through a lot of time and effort and money to extract him from the Kingdom and from Veeby.

"Speaking of brands and such," Jacob asked, "What about my ex-wife?"

"She'll get a report with our condolences."

"Are there telephones in your jail?"

The question seemed to confuse Handy. "Telephones?" he repeated. "I suppose so, but you wouldn't call her would you?"

"No," Jacob readily agreed. "That would put the law right back onto me. Wouldn't it?"

"Most certainly," Handy said. "There are many advantages to being dead. You name is deleted from all of the databases. Can we rely on you?"

"Count on it," Jacob said.

"We thought you were working with the Kamsi. We traced your telephone calls. They all seemed to originate in the Middle East."

"I know," Jacob said. "My wife lied about my whereabouts."

"We contacted her repeatedly trying to find you."

"She had her ways."

"With help from the Chinese," Handy replied. "Hell, with a few dollars you can buy damn near anything including fake phone numbers."

Jacob turned away feeling a rush of pain and gratitude that there were people in the world that actually valued him as a scientist and as a person. Walking a couple steps toward the bunk beds Handy stopped him and brushed his hands against Jacob's trousers.

"Bug check," Handy said not seeming to see Jacob's distress. He looked Jacob's clothing over carefully. "Centipedes," he explained. "And scorpions."

Jacob broke into a cold sweat, because even with his years of working in genetic engineering, with all his experience in agriculture and plants, he still hated some bugs.

"And let's see," Handy added, "We need to make sure other critters didn't follow us inside."

"I like bugs," Jacob said, "except ones that bite."

"They all bite around here," Handy said, "but it's the snakes around here that kill."

"I *hate* snakes!" Jacob said. He took a deep breath.

"Well I guess we both made it to our middle years," Jacob said philosophically.

"Ain't it the truth," Handy said then gestured to a refrigerator. "There's a refrigerator over there. Help yourself. Sack out for a while. We'll be leaving soon."

Jacob wandered around the bunkhouse. He crawled onto a bunk fully dressed and went to sleep. He slept like a baby in his mother's arms. Sweet sleep. Without a care in the world.

After a while he thought that Handy said something to him, but his exhausted body was too weak to wake up. The pleasant dreams faded and then were replaced with a dream that somebody was examining his feet again.

Jacob moved and might have grunted as a needle pricked his skin.

"Blood," a man said.

"Put a band aid on it," another male voice said.

The stick of the needle vanished almost instantly and Jacob fell into a deep sleep. Everything was wonderful except his dreams changed.

16

Jacob's mind sank into a hole filled with images. He dreamt the same dream over and over that a small aircraft was flying low and fast. The airplane narrowly missed the trees.

The pilot was a kid that obviously loved to fly. Jacob knew that, because the kid kept laughing. Every time the aircraft missed a rock or a tree the kid let out a whoop.

The aircraft dodged tall cactus at the last instant. Scrub trees with branches loaded to bending with the weight of vultures roared passed the aircraft's windows. Jacob dreamt that his eyes were open and that he could talk to the pilot and to Handy who was in the back seat. In his dream Jacob couldn't understand why Handy seemed bored.

Jacob saw himself in the passenger seat as if he were separate from himself. In his dream he kept asking the same question, "Where are we going?" and the pilot kept giving the same answer, "To the station. Now shut up!"

After a time the airplane landed in the desert. The pilot kept the engine running. He talked incessantly on the radio apparently waiting for permission to continue the trip. The daylight faded and the stars came out. Then, deep in the night, the airplane took off again and flew under a moonlit sky. Jacob watched a limitless expanse of landscape illuminated by the moon's silvery light. He could even make out the vultures taking wing as the airplane skimmed the treetops. Black wings fluttered against a sparkling desert floor. Ahead of the airplane he saw a small cross on the ground. It had electrically lighted runways.

"What city is that?" Jacob asked.

"That's the Station," the pilot said.

"The Station?"

"Will you shut your mouth?" the pilot asked. "You'll find out soon enough. Okay?"

"Just curious," Jacob said innocently.

He watched the cross enlarge and finally came up to meet the airplane. When the wheels touched ground the aircraft rolled to a stop Handy reached across the seat to get Jacob out of his seat belt.

"Can you get out?"

Jacob fumbled with the door latch. Hot air filled his lungs as he stepped out onto the aircraft's wing. He put his hand to his forehead. "My head hurts."

"We'll take care of your head," Handy said scrambling out of the aircraft after Jacob.

They walked away from the idling aircraft as, behind them, the pilot revved the engine and the airplane started back down the runway. Jacob wanted to watch it depart, but Handy pulled him toward the edge of the tarmac and then into a huge tin hut with military personnel milling around. An officer came up.

"Delivery?" the officer asked.

"Ebbtide. This completes the set."

The Officer's face settled into a grin. "Ooh," he drawled, "Ooh that's real good. Hot damn! That's real good. We finally got him. Ooh, damn that's good."

"Yeah we finally got him," Handy said.

He led Jacob through a busy office. Old typewriters clacked and modern computer printers buzzed. They went down a hallway and outside, passed a small garden area, and then around a large greenhouse and entered another building.

Jacob followed Handy into a small living unit.

"You'll have this space," Handy said. He flipped on a light and Jacob was surprised to see the overhead lights actually work. On a small dresser he saw a TV set. Jacob pointed.

"Does that work?"

"Sure."

When Handy walked out closing the door behind him Jacob saw a sign on the back of the door, 'Welcome to Groom Lake.' There were photographs on the knotty pine walls of aircraft and rockets.

The windows had blue curtains with a pattern of little white ducks paddling this way and that. A bathroom off the bedroom had white tile floor and walls. It was the plainest bathroom with the oldest pull chain toilet Jacob had ever seen.

As he stood in the doorway the whole bathroom seemed a throwback to the 1950's. A medicine cabinet had a bottle of aspirin. Jacob took two, walked back out to the bedroom and turned on the TV.

The set was tuned to one military channel, but no sound. The same video images of the greenhouse and the tiny garden plot were the sole programming. After a few minutes of watching the camera pan over the greenhouse Jacob flipped off the set.

He stripped out of his dirty clothes took a long hot shower and idly thought that he might rack up more prison time by using so much hot water. He'd gotten used to the two-minute showers at the Hines Prison House, but at this point he didn't care.

After thirty minutes of letting the water run he felt guilty, dried off and walked into the bedroom. He found new socks and clothes in the closet, although the clothes were very limited in style. All garments were one-piece jumpsuits, zipping down the front with a hatch in the back for doing business.

"Now," he said, "I'm all dressed up and no place to go."

He sat on the edge of the bed glancing around, from the photos, to the TV, to the sign on the door. He got up and opened it a crack. A military policeman faced him. The solider clicked his boots German style. Jacob closed the door.

"Checking and testing," Jacob said.

He heard the trooper click his heels again.

"Do you have a brother named Adolph?" Jacob asked and heard the soldier click his heels twice.

Jacob moved quietly to a window and slowly pulled open the curtains. The window was barred. He peeked out and saw the silhouette of another soldier standing guard duty.

A moment later Jacob mused that, all in all, as W. C. Fields an old time actor had engraved on his tombstone, he'd rather be in Philadelphia, but still this *was* better than Stonegate East.

His fingers ran up and down the silk of the new jumpsuit. It was warm, but not hot. The cloth was very soft. The cuffs and wrists had elastic bands that formed gently, but securely around the ankles and wrists. He snapped the bands a few times to test the stretch.

As he sat on the bed he realized that he was a happy man. All he needed was a new jumpsuit, hot shower and lights that worked. Under the bed he found a pair of black felt-lined boots that happened to fit him perfectly.

The boots zipped up the side of the ankle. He wiggled his toes and then decided to test the bed. Lying over, fully dressed with his boots on, Jacob stretched out flat on the bed. Even the pillow seemed fitted for him. He felt like royalty. T*his* was living.

A knock at the door brought Jacob to his feet. "Guard?" Jacob asked.

"Just me again," Handy said. "Feel up to a tour of the facility?"

"Now?"

"No time like the present right? Besides," Handy's voice took on a cautious tone, "You've got some decisions to make. No time like the present. Huh?"

"By all means," Jacob said stepping out into the hall and closing the door behind him. In a few moments Handy took Jacob around a collection of military barracks style buildings to a large structure.

Jacob stopped and looked at it gleaming in the moonlight. It seemed to radiate heat; Jacob could feel it even through his jumpsuit. Even the air above it was heated. The whole thing seemed to radiate energy. Handy walked to it.

"I wanted to show this to you tonight," he said. "Time is critical. I'll need your answer tomorrow morning."

"Answer?" Jacob asked. "What's the question?"

"We won't keep you here if you don't want to stay. See, if you want to go, well," Handy let his sentence go unfinished. "The Government has been looking for you since you dropped off the edge of the world in Israel."

"Dropped off the edge of the world? What are you talking about? Leviman knocked me unconscious and shipped me like so much meat to the Kingdom!"

"Leviman is a man in the middle," Handy said. "He has his own problems."

"He betrays his own people!" Jacob said emphatically.

Handy stopped, reaching out, taking Jacob by the arm. "Leviman?" he asked. "We have no information that he's betrayed Israel."

"He sent me packing!"

"We had information that you were working for the Kamsi. I think Leviman had the same information. It seemed like good information at the time. We didn't want to believe it, but…"

"Disinformation," Jacob said. "I can't see how anyone would doubt me," but when Handy grunted, Jacob felt a tinge of embarrassment. He shrugged. "Well," he said by way of further explanation, "I can see how some people might get the wrong impression."

"You *do* have a history of passing information for money."

"*Only to Israel!*"

"Yes," Handy said, "But that distinction is lost on people. Anyway we wanted to speak with you directly. As you say false information. Getting everything second hand, well, it's hard to make good decisions."

"I'm grateful," Jacob said. "Don't misunderstand. This is much better than what I've gotten used to."

"Living here is hard," Handy replied, "but we are free. With your recent experiences that might matter to you."

"I can go if I wish?" Jacob asked.

"You can yes," Handy answered, "but first we are willing to show our good faith by disclosing our new energy system to you. We'll keep nothing back from you, but we won't keep you here if you want to leave. Fair enough?"

Handy walked to the structure. Grass sloped sharply up its massive sides. The structure reminded Jacob of a truncated pyramid. It had four sides slanting up, but flattened out on top. Two external vertical columns rose up to the top of the structure. An enclosed bridge went from the top of the elevator to the top of the pyramid.

"Elevators," Handy said. "The structure is a huge tank of heated water."

"Where'd you get the water?"

"This water was shipped in, but if this system works we can use seawater or brackish water."

"What?"

"Solbean," Handy said as he walked to one of the elevators. "We can mix and treat the bad water with a small amount of…"

"Yeah," Jacob said. "The Kamsi are doing the same thing. I'll explain it later…"

Handy walked into an elevator. They rode it to the top. "Stairs?" he asked

"It's alright," Handy assured him, "This device is a power station. We're in a power station. We won't get stuck. This structure generates electricity."

They came out at the top and walked across the bridge to the flat top of the structure. Jacob saw that the flared out area on top of the structure was hollow and contained a dark liquid.

They stood on a ledge that extended all the way around the structure. "Is that water?" Jacob asked pointing to the liquid filling the interior.

"It fills most of the power plant."

All around the square enclosure the slanting walls were shinny ceramic tiles with protrusions here and there for rubber hoses. Steam hissed out of unseen ports.

"An energy source for the future," Handy said. He pointed to the star-filled heavens far above. The Milky Way spread out in a vast field of stars.

Jacob's practiced eye searched for satellites, but he could see nothing moving against the starry sky.

"We have an active space program Doctor."

"Do you launch from here?" Jacob asked.

"Oh no we don't launch," Handy replied. "It's all done from China. They operate the whole system."

"China?" Jacob asked. He let his gaze slowly descend from the sky back to the huge structure with the dark water shimmering inside of it.

"The Chinese build and launch rockets to our specifications. The world thinks that they are launching weather satellites, but our rockets are much more than that.

"Our satellites are large solar collectors that generate photovoltaic current. The current powers a laser transmitter. The system beams energy to the -Power Station."

Jacob followed Handy's gaze down to the water. So, the Americans were using high power lasers to heat water, but to what benefit?

Handy gestured up and then down to the black water in the enclosure. "It's salt brine. It boils at a very high temperature and it holds the heat."

"Water holds heat?"

"The sides of the structure are insulated. Escaping heat helps the Solbean grow.""

Jacob let his gaze come down from the starts slowly to the salt pool. So, he thought, the Americans have an answer for Solbean. They're chosen to combine a satellite solar collector with saltwater. The microwave radiation from space heats the water molecules the same as in a microwave oven. The salt in the water raise the water temperature and hold the heat when the satellite isn't overhead. With the heat they generate steam and with the steam they run the turbines that generates electricity.

"Does it work?" Jacob asked.

"Yes, quite well," Handy said, "but of course it's not the whole answer. We still need liquid fuels to operate machinery such as farm tractors."

"Yes," Jacob said, "but," he shook his head. "It must be very expensive electricity."

"One of our satellites can keep the salt brine heated in over fifty structures. Actuate Global Positioning Systems are the answer.

"Each structure generates enough electricity to power five thousand homes. You do the math. So yes, the electricity is more expensive than it used to be, but now it's reliable. Industry and people can be confident of the price. Reliability you see?"

"Yes," Jacob said and he didn't say so, but understood that's why they needed him and scientists like him, to do genetic research on developing new strains of Solbean.

Solbean, the miracle plant that could grow with saltwater— or salt brine. Jacob focused on the pyramid structure and nodded in quick understanding. They were in a desert. The greenhouse they passed was a research facility. The Americans were planning a new energy system.

"You are planting Solbean?"

"Yes. Of course. It's planted on the sloping sides of these structures."

"How do you harvest?"

"The Solbean is planted on the south and west walls of these and other structures. A machine travels around the perimeter harvesting the Solbean mechanically. Naturally, the plants are very close together, very dense. They are irrigated from inside the structure. We tap solar energy on the west and south walls to boost our thermal input to the system.

"The hot salt brine is a source of warmth and water for the growing plant. With this system we can grow Solbean that is adaptable to very hot conditions in any climate including Alaska. We are also moving away from petroleum based oil."

Jacob extended his hand out. As they shook hands Jacob gestured out to the structure. "You're a man after my own heart. You have some good engineering going for you."

"We have good people," Handy replied. "And we'd be glad to include you on our staff."

"I've been looking for a chance to do research again," Jacob said. "I'm grateful."

"You're probably also tired. That drug I gave you was a precaution. You have a reputation of being troublesome. Don't take offense."

"I can't imagine why anybody would feel that way about me," Jacob replied innocently. "Really. I'm only a scientist. My life is my work, but I understand. Taking a flight in a small aircraft you want to insure nobody grabs the controls." "I understand," Jacob said again. "Better safe than sorry."

They walked along the interior ledge of the structure until they came to the elevator entrance.

"Can you tell me about the Centas?" Jacob asked. Handy shook his head.

"No, I can't," he said, "because I don't know. I know the Kamsi have trouble with them."

"Is my daughter somehow involved in that?"

"Again I don't know," Handy replied.

"Any idea where she is?" Jacob asked. "I'm worried about her."

"About that I can't say," Handy said. "Others may know."

They walked on in the evening for a few minutes. Jacob smelled food. Ahead of them, in one of the tin-roofed huts a cafeteria was doing a brisk business.

"Let's find Herman," Handy said.

"Herman?"

"Burger," Handy said. When Jacob made a startled sound he looked over. "He and Zada are here," he said and then paused staring at Jacob. "And," Handy let the sentence go unfinished.

"Here?" Jacob asked. "For how long?"

"They came directly from Israel with agreement from Leviman."

"Here!" Jacob shouted. "I've been worried sick about them for two years!"

Handy stopped again. "Doctor," Handy said with his voice low. "We relied on Leviman to keep us informed. This is a high level project. The UR President carries messages in a diplomatic pouch. We can't trust electronic communication.

Everybody is spying on everybody. Everybody has software now to break the codes."

"I've broken a few codes in my day," Jacob said, "but Holy Moses," he shouted, "I should have been told!"

"We certainly would have told you," Handy said contritely, "but of course for most of this year we thought you were in the Kingdom taking money and trading secrets.

"Then we heard that you and another man had skipped out of Stonegate. That was news. We didn't even now you were in the country."

"But I told you!" Jacob shouted. "I told you I was in the Hines Prison House!"

"But we followed up on it the best we could," Handy insisted. "If we had known we'd have come and got you!"

"The Kamsi are clever people," Jacob admitted.

"And of course," Handy said, "There was also the money."

"What money?" Jacob asked.

"Money in a Swiss bank account. A numbered account in your name. Money was put into your account every single month by the Kamsi."

"I never saw a penny of it," Jacob said, "because I wasn't in the Kingdom. I was stuck in prison!"

"We know that now," Handy said. "Those Kamsi are clever bastards."

"So is my ex," Jacob said.

"We contacted her four or five times," Handy explained, "But always came away convinced that you were lying."

"Between my ex and the Kamsi I didn't have a chance," Jacob admitted. "I don't find fault with you. I know you tried."

"Yes, we followed the money," Handy said.

Jacob nodded. It was always about money

"When did you know?" Jacob asked.

"When Gilla contacted us we knew for a certainty that you were coming back to the UR. Of course we paid his demands." He stopped again taking hold of Jacob's arm.

"For the record," Handy asked, "Gilla said your contract buyout was over two million gold dinars. He screwed us didn't he?"

Jacob swallowed hard and gave Handy a sheepish glance. "You were in hock to them for two mil?" Handy asked incredulously. "How is that possible?"

Jacob squared his shoulders and walked away. "It's a long story," he said over his shoulder.

"In here," Handy said, catching up and walking by a long low building. They walked into the cafeteria and waited patiently in a long line of military personnel. The cafeteria was packed with officers and enlisted people of all ranks, but most seemed to be in the Air Force.

Handy ate a late breakfast and poured heavy maple syrup over pancakes and the sight of that caramel brown liquid flowing so freely made Jacob's mouth water.

Jacob also ordered pancakes with coffee and eggs and sat with Handy and two others. Jacob ate silently letting the hot tasty food roll over his tongue. He had forgotten what pleasure there was in the simple things: pleasant company, hot food, electric lights that work. Even the drug Handy had given him hours before put him in a happy frame of mind. He liked these people and this place. After eating Handy excused himself from the table.

Jacob finished his meal and was ready to leave when somebody he recognized came into the dining room. Jacob knew that portly gait and the stock of white hair.

"Burger?" Jacob asked startled. "Is that you?"

"Jacob!" Herman sat down and then got up again. "What was that?" He pointed to Jacob's empty plate.

"Pancakes."

"I'll be right back."

Herman waddled off to the serving counter. He came back with a plate stacked with food. Jacob relaxed and let his old friend eat.

"Is Zada here?" Jacob asked when Herman stopped eating.

"Oh sure," Herman said. "He sleeps all day and works at night."

He looked across the table and gave Jacob a mischievous look. "I think he's still trying to figure out how in hell you broke our Code!"

"Ahh!" Jacob said giving Burger an off-handed gesture, "That was child's play, but your program has real possibilities."

Herman explained to Jacob that the same satellites that generated laser energy for the power system were also used to send and receive laser communication-streams of high-energy light.

The Kamsi were looking for a conventional ground station sending *electronic* messages to control their fleet of Centas-not laser light communications. In fact, unless they were within a line of sight and couldn't actually detect the laser signal, there was no way for them to find out. Laser communication is much more secure than electronic signals. The Kamsi were looking for the wrong type of signal.

"So it's the Americans that are sabotaging their system?"

"It certainly is."

"I understand?" Jacob said. "Laser light isn't electronic. The Centas communication system is all electronic. Not light sensitive."

Herman smiled as Handy came to the table.

"No?" Herman asked.

"No," Jacob said. "Electronic devices won't pick up laser light signals."

"The Kamsi are not a manufacturing economy," Herman explained, "They are buying most of their components from China." His smile deepened.

"My god," Jacob let air out of his lungs. "You've paid off the Chinese?"

"They need energy too," Herman said, "And remember it's still a global economy. The President and Congress are firmly committed to it."

"I think they should be committed," Jacob replied angrily, "but what the hell…"

Herman pushed his empty plates aside. He leaned over the table. "It's negotiation that's all, Jacob. The Kamsi knows it's us, but of course we can't, for the sake of legality, say so.

"Civilization depends on energy, you know, and it also depends on the *price* of energy. We are in a tough game with the Kamsi to keep the price of energy low. We can't let them monopolize agricultural fuels as they did for years with petroleum."

Herman looked hungrily over at the food display and made ready to hoist his mass up onto his feet. "It's price," he said. "The Kamsi have millions of square kilometers! They'd drive the price down and our political leaders would put us right back as price slaves! We'd be importing biofuel. Same as oil!"

"Yes," Jacob said. "And when we destroyed our own energy production the Kamsi would raise price."

"Honest to God," Herman said moving away from the table, "Our corporate masters never learn!"

"It's a global economy," Jacob said without conviction. "People should not be dependent on strangers for their survival! If we let the Kamsi dominate the market we'll be right back where we were with oil.

"I was in the Kingdom," Jacob added. "I can tell you that you are driving them crazy."

"Good," Herman said.

When Herman waddled to the serving line again Jacob suddenly felt very tired, almost weepy with exhaustion. He decided his education about the Americans could wait for another day.

Handy got up and was waiting for Jacob as he left the cafeteria. He walked Jacob to his room. They shook hands again and Jacob watched Handy leave the barracks.

That night Jacob slept with the light on. When sunlight broke through the curtains the following morning Jacob put his hands behind his head and decided to lay there, flat on his back, until somebody came along with breakfast. In a few minutes the door opened. Handy stood there empty handed.

"No breakfast?" Jacob asked.

"Ready?"

Jacob rolled off the bed, reached sleepily for his boots, but realized they were still on his feet and slowly stood up.

"So?" he asked. "What do we do first?"

Handy walked into the room with two guards.

"Do we need to restrain you?"

Jacob was momentarily stunned until he saw that Handy was smiling. "Take me to breakfast," Jacob said, "And all will be well."

"First I need an answer," Handy said.

"Yes, of course, I want to stay," Jacob said. "Where else would I go? What would I do? Push a lawn mower for Gladys?"

Handy scratched his head. "I'll take that for a yes," he replied.

"Can I have more pancakes?" Jacob asked.

"I'll do better than that," Handy said. "Now that you're staying. There's someone here who has been worried about you."

He stepped aside and Jacob saw someone else standing in the doorway. He rubbed his eyes. Then he rubbed his eyes again. The girl was slim and attractive. Her hair was done up in a certain way.

"Hi Daddy," Talya said. She walked into the room. Jacob sat down heavily on the bed, his jaw open and only when he slowly stood up did he find his voice.

"How long have you been here?"

"About two years."

When he held out his arms she slowly came forward and let him give her a hug. When he released her he sensed her uncertainty. She was nervous.

"Talya," he said softly while brushing her hair, "Why the long face?"

"Leviman was in the process of sending you and me together when that situation popped up with Dexter at the water irrigation plant."

"Leviman had last minute doubts about me?"

"I'm sorry Daddy." She sat down by him. The door closed softly behind Handy as he walked out to the hall. Talya sat with her arm around Jacob's shoulders.

"Daddy," she asked, "Mister Handy said that you were in mother's jail? Is that right?"

"But I told him!" Jacob insisted

"I'm sorry," she said, "but Mother denies it! I talk to her once a week. And there was no record of your incarceration in Boston. No record and we traced your phone calls to the Kingdom. We were convinced you were calling from the Middle East!"

Jacob looked tearfully over at his daughter. "I know," he said softly. "The Chinese play a good game."

"So do the Kamsi."

"Yes," Jacob said. It was all too clear to him. Everybody has an agenda.

"We tried to confirm your story," Talya said, "but all the facts simply pointed to…"

"That I was lying to cover my activities in the Kingdom?"

"Yes," Talya said, "The Government didn't know you were in the UR until Doctor Veeby filed a report that you had escaped from Stonegate. Nobody believed you were in the country. Mother denied it and even sent fake reports to prove that you were lying."

"But you could have sent investigators!"

"We did!"

"And?"

"I think they were bought off," she said sadly, "or else mother conned them. She has resources."

"God!" Jacob shouted. "How could she hate me so much?"

"The Government always wanted you here. They would have come for you, but of course between one thing and another. And, of course, Father, you *did* arrive from the Middle East."

"I'm sorry," she said softly. "Leviman organized the whole thing. It was run like a military operation. Nothing could be left to chance.

"Stay here Daddy," she pleaded. "It's a good place."
"Yes. I saw Herman last night."

"The Americans are putting together a research team. You were always high on their list. They couldn't figure out why you were working in the Kingdom."

Talya's comment caused Jacob to gulp air in sudden disgust. His eyes bulged. Tears ran down his face. He turned to his daughter taking her by the shoulders.

"Why I was working in the Kingdom!" he shouted.
"God Damnit!" he bellowed, "I was running to avoid capture! That bitch mother of yours had me bouncing like a rubber ball!"

Talya wrapped her arms around him. She held him for a long moment then looked up at him.

"Daddy I'm so sorry. Mother said you were in the Kingdom. Everything seemed to prove her right. None of us knew you were stuck. Not until you slipped out of Stonegate. Oh Dad. You escaped!"

"I was alone," Jacob said feeling self-pity. He straightened up. "Well," he said, "Now the only loose end is Leviman."

"He's with us," Talya insisted.
"Another bastard," Jacob swore. "I think he's working for the Kamsi," but the words were no sooner out then he met Talya's gaze.

"Leviman received information that you had gone over to the Kamsi. You put American satellites out of commission for six months. You used Kamsi money to escape from Stonegate. What was he to believe?"

Jacob saw her eyes watering. She always wrinkled her nose when she cried.

"Talya," he said softly, "the crop was dying! What was I to believe? You tell me, Talya, with the whole damn crop dead! What was I to believe?"

"Oh Daddy," Talya sniffled, "You're so smart in some ways," she hesitated and reached for his hand. "Jacob," she said in a more business like voice, "It wasn't the American satellites that killed the Solbean! It was Dexter at the Central Water Plant. He injected a substance into the water supply. The usual safeguards wouldn't detect it."

"I heard about Dexter."

"He was your man!" Talya said, "And then things started going badly with the American satellites."

"And the Kamsi spread rumors about me," Jacob said. "I can see how Leviman got the wrong idea."

Talya hesitated and added, "Leviman had Dexter shot," and she looked fearfully up to her father. "You were lucky to run when you did," she said, "because the Kamsi gave him credible evidence that you and Dexter were working together and of course…"

"But I didn't run!" Jacob shouted. "Leviman handed me over. His thugs hit me on the head. I still have headaches!"

"He was testing you," Talya said, "There was evidence against you, but Leviman knows you. He knows you spied for Israel. He just couldn't bring himself to believe that you would betray your country."

Jacob slumped over onto the sink twisting the towel trying to control his emotions, trying to understand how the President of Israel could turn him over to his blood enemies. How Leviman could possibly think that he would work for the Kamsi. Then, the thread of logic straightened itself out.

"God," Jacob said realizing for the first time that his sabotage of the American satellites was the 'frosting on the cake.'

By going around behind Leviman's back, by throwing a monkey wrench into those two satellites, he had unwitting convinced Leviman that the Kamsi information about him was correct. That he had, in fact, been working with Dexter.

His sabotage of the American satellites would have only convinced Leviman that he was acting with the Kamsi to meet 'fire with fire.'

Jacob cringed, realizing how his actions would look to Leviman. "God," Jacob said again. "I'm so sorry!"

"It never was the American satellites," Talya said.

Jacob leaned over the sink, washed up, dried his hands and let Talya lead him out of the bathroom. In the hallway they found Handy waiting for them.

"Are things cleared up?"

"Yes," Jacob said. "I've been a fool."

"It's the world we are living in," Handy said. "It makes people crazy."

"People like me and my ex-wife have a head start."
Jacob put an arm around Talya's shoulder as they started down the hall, but then Jacob remembered something.

"What about you?" Jacob asked Talya. "The Kamsi seem to think that you and the Americans are behind their Centas difficulties."

"Yes," Handy replied, "We used Talya as leverage against you. You know. We planted false information to get them to doubt your sincerity. We tried to use her to give you more trouble, you know, thinking that you were working for them."

The Americans had piled more trouble onto his difficulties by getting the Kamsi to think that he had defected to the Kingdom as a 'mole!' No wonder they hated Jacob's guts!

"Damnit!" Jacob cussed loudly.

Still, all in all, he was truly happy with the prospect of doing research again and maybe now, if he could control his temper, he could make things right.

End

Selected other fiction by the same author

War of the Lights

The Sunburst renegotiation

The Weed Farm

Bled White

Let's make a deal

Shell game

Earth, mind and murder

Hidden Jungle

The Balloons

The Deep and the Dark

Split personality

White shadow

Becky's flight

Super lube

Nothing can go wrong

Super sleuth

The falling stars

Fiction continued

Navigans

Power play

Three novellas

The Black parabola

Angel dust

Dreaming

Triple threat

China moon

Drug Club

A submarine called Blowfish

American parallax

Creative nonfiction

Unconventional Transportation

Unconventional transportation and energy

Unconventional Marine Carriers

Maritime Biofuel Production

Alternative Energy and Synthetic gasoline

Alternative Energy as fact or fantasy

Return of the tall ships

Sea power and solar energy

For a complete list of work please see Amazon. com or Kdp. com.

Brief author bio:

Robert Vollmerhausen worked as a patent draftsman and as a drafting supervisor in industry and government for approximately thirty years.

He holds five United States patents on transportation technology including the mechanical aspects described in his work *Unconventional Transportation*.

He is a Viet Nam veteran and a graduate of Wayne State University in Detroit, Michigan. He is now retired and lives in southwestern Virginia.

He may be reached at Ceninven@gmail.com

www.ingramcontent.com/pod-product-compliance
Lightning Source LLC
Chambersburg PA
CBHW051443170526
45166CB00001B/90